P9-DMD-712

The Enchanted Braid

The Enchanted Braid

Coming to Terms with Nature on the Coral Reef

OSHA GRAY DAVIDSON

John Wiley & Sons, Inc.
New York • Chichester • Weinheim • Brisbane • Singapore • Toronto

This book is printed on acid-free paper. ♾

Published by John Wiley & Sons, Inc.
Published simultaneously in Canada.

Library of Congress Cataloging-in-Publication Data

Davidson, Osha Gray.
The enchanted braid : coming to terms with nature on the coral reef / Osha Gray Davidson.
p. cm.
Includes bibliographical references (p.) and index.
ISBN 0-471-17727-X (cloth : alk. paper)
1. Coral reef biology. 2. Coral reefs and islands. 3. Human ecology. 4. Endangered ecosystems. I. Title.
QH95.8.D38 1998
578.77'89—dc21 97-39797

Printed in the United States of America.

10 9 8 7 6 5 4 3 2 1

In memory of Theresa Lynn Riffe

. . .

I truly believe that we in this generation must come to terms with nature, and I think we're challenged as mankind has never been challenged before to prove our maturity and our mastery, not of nature, but of ourselves.

—Rachel Carson

Contents

Preface xi

PART ONE The Soul of the Sea

1 "Who Has Known the Ocean?" 3
2 Animal, Mineral, Vegetable 13
3 Darwin in Paradise 23
4 The Rise of Corals 37
5 The Heart of Lightness 51
6 The Outer Strands 63
7 A Song of Love and Death 73
8 Fish Stories 85
9 Neither Brethren nor Underlings 99

PART TWO Human/Nature

10 The Jakarta Scenario 115
11 "Either We Go Deep or We Starve" 129
12 The Apo Scenario 147
13 Return to Oceanus 159
14 Disasters, Catastrophes, and Tragedies 177

15 Unweaving the Outer Braids 195
16 Once More to the Keys 207
Notes 225
Bibliography 241
Acknowledgments 255
Index 259

Preface

Back in the early 1970s, when I was living the life of a beach bum in Key West, I'd go down to AT&T beach at night, stretch out upon the still-warm sand, and gaze at the multicolored lights flashing far out at sea. I knew that the flickering lights marked the coral reef five miles offshore, but that was the extent of my knowledge then. During the day, I'd put in my full eight hours on the beach, snorkeling and drowsing in the sun, hanging out with friends. There is an Italian expression for this sort of existence, although I didn't learn it until much later: *dolce far niente.* It means, literally, "sweet doing nothing," and that pretty much sums up life in Key West back then.

Coming from Iowa, I was for a time largely content to snorkel among the crumbling pilings of the old pier near Vernon Street, where schools of fishes darted in and out of concrete blocks, and among the small patch reefs beyond the seagrass beds. But still I wondered what lay beneath the water marked by the lights. Mysteries feed on time, and as the months went by, the beacons became for me what the light at the end of Daisy's pier was to Jay Gatsby—the symbol of an obsession. Finally, I talked my friend Cory into taking me out in a borrowed boat. It was not the best weather for such an excursion. A tropical storm was caroming around the Caribbean, and the water out at the reef was dotted with whitecaps. At one point we saw four separate waterspouts writhing like immense snakes between the sea and the leaden sky. I was having second thoughts about the trip, but Cory was a Conch, a Key West native, and so by definition a bit crazy. He wasn't about to let a few waterspouts spoil our fun.

"Nothing to worry about," he yelled as I kept an eye on the waterspouts. So I convinced myself that I wasn't afraid—not much of a feat when you're young—and dived into the water.

There is another phrase, much used today, that applied to what I experienced that afternoon on the reef: "information overload." I actually remember very little of what I saw on that first trip. There was simply too much to see, and none of it fit into the world I had known until then. But I vividly recall the *feeling* that I had somehow stumbled into a different universe, one constructed of bright colors and populated by exquisite and bizarre creatures.

I have returned many times over the past two decades, and the thrill of the reef remains just as overpowering, and the miracle of life there just as ineffable, as it was back then. It is this sense of wonder and magic on the reef that I want most to convey—that and the fact that this unique and enigmatic underwater realm may be fading from our world.

What follows isn't a scholarly treatise, but a natural history of coral reefs, and of humanity's relationship to them, written by a layperson for laypeople. Whether the work turns out to be an ode or an elegy is still very much an open question, and the answer lies—where else?—in our hands.

The Enchanted Braid

PART ONE

The Soul of the Sea

All the beasts that roam the earth and all the birds that wing their flight are but communities like your own. They shall all be gathered before the Lord.

—The Koran

1

"Who Has Known the Ocean?"

In the end we will conserve only what we love; we will love only what we understand; and we will understand only what we are taught.

—Baba Dioum

There is a story about an ecologist who thought he had found a way to decipher the coral reef. He believed that all he needed to do was to feed the known data about reefs into a huge supercomputer, type a few commands, push the ENTER key, and out would come the secrets of life on the reef. He gave it a whirl. What issued from the computer was not the reef's trophic pyramid, the intricately drawn food web that the scientist had dreamed of, but billowing clouds of smoke.

Presented with such an impossible task, the supercomputer had blown up.

The smoke is no doubt a touch of melodrama, a cinematic embellishment. (I've had several computers die in my arms, as it were, and there has never been so much as a wisp of smoke.) The entire story, in fact, sounds like just one more cautionary tale about the dangers of scientific hubris.

But knowing several scientists, and a bit about computers and their limits, I'm inclined to believe the general facts of the story. More important, having been around reefs for many years, first as a beach bum in the 1970s, and over the past few years as a more serious student of this most complex of all marine ecosystems, I am convinced of the tale's essential validity.

Rachel Carson wrote many true and important things in her productive life, but nothing truer nor more important than this: "Who has known the ocean? Neither you nor I . . ."[1] It is surely ignorance that defines our intellectual relationship with the ocean, and even that iridescent sliver of the sea called the coral reef remains a mystery to us, even though most reefs are close to shore and therefore by definition in shallow and accessible waters.

The source of our ignorance is twofold: First, we are terrestrial beings and our study of life largely ends at the water's edge, even though the oceans cover 71 percent of the Earth's surface.[2] Second, there is too much life on the coral reef ever to think that we know it fully.

Consider the ocean itself. Nothing on land prepares you for the dimensions of the world beneath the water. The deepest valleys on dry land become trifling nicks on the ocean floor. If engineers could build a structure as tall as the Grand Canyon is deep, six of these monoliths could be dropped into the Pacific Ocean's Mariana Trench, one on top of the other—and the pile would still be nearly a mile from the water's surface. Think of the Gulf Stream, the river of warm Caribbean water flowing north and east within the Atlantic Ocean. If you stood on the shoreline in the Canadian Maritimes facing out to sea, more Gulf Stream water would slip silently by you in one second than is contained in all the roaring rivers on Earth.[3]

But the throng of life found in the ocean is even more impressive than its grand physical structures and sweeping currents. The sea is mother to all life on the planet. This lineage is most apparent in animals. Of the thirty-three animal phyla (or major divisions of the animal kingdom), thirty-two are found in the sea, and only eleven of these are also found on land.[4] There is a single animal phylum that is exclusively terrestrial (the Onychorphora, a small grouping of segmented wormlike organisms that live in tropical forests).[5] This biodiversity makes sense because the oceans constitute 99.5 percent of the planet's biosphere, or living space.[6] The oceans have had not just space but time working in their favor. Around 3.5 billion years ago, while the young Earth's atmosphere was still inimical to life, primitive bacteria formed in the ocean's waters.[7] Life in the sea had plenty of time to evolve into myriad forms—algae, seaweeds, jellyfish, sea urchins, fishes, all the major vertebrate groups, and, of course, corals—before the very simplest land plants first appeared around 400 million years ago.[8]

There is no corner of the ocean that is alien to life, although early scientists believed that life existed only in the upper reaches of the oceans, where light penetrated—the so-called photic zone. The British explorer Sir John Ross rocked the scientific world in 1818 when he hauled to the surface living organisms from the Arctic seabed over a mile down.[9] Today, researchers are still discovering new marine life forms at ever-increasing depths. In the summer of 1996, scientists announced that they had sequenced all the genes from a microorganism thriving several miles below the surface in the mineral-laden boiling water spewing from hydrothermal vents.[10]

But although it is true that life exists everywhere in the ocean, nowhere is it so varied and so dynamic as on the coral reef. In comparison with the reef, the rest of the ocean is a desert. Although coral reefs represent less than two-tenths of 1 percent of the area of the global ocean,[11] approximately one-third of all marine fish species are found in this tiny zone.[12] Even more remarkable, coral reefs are home to approximately one-quarter of *all* marine species.[13]

Coral reefs are literally overflowing with life; wherever you look on the reef, you will find life in astonishing variety and abundance.

One example: On a single coral reef surrounding one tiny Australian island, there are one thousand known species of fishes.[14] Zoom in closer: a scientist has counted 620 species of shrimp living on corals.[15] Get even closer; go inside the coral: there, searching through a labyrinth of passageways within a single colony, an investigator found 103 separate species of a single kind of worm.[16] Perusing scientific journals, you can find dozens of equally compelling examples. But, for the sake of brevity, perhaps it's better to just sum up: The reef is home to a spectacular yet unknown number of species of fishes, shrimps, worms, snails, crabs, lobsters, sea cucumbers, sea stars, urchins, anemones, sea squirts, and sea plants—not to mention several hundred species of corals. The *lowest* scientific estimate is that reefs are home to about one million individual species. For the upper-end estimate, multiply that number by nine.[17]

The only terrestrial analog to the coral reef is the tropical rain forest. And, in fact, coral reefs are often referred to as "the rain forests of the sea." But while rain forests outdo coral reefs in sheer numbers of species, reefs contain far more phyla than do rain forests.[18] And bear in mind that rain forests cover twenty times more area than coral reefs.[19]

If we were not so terrestrial in our thinking, we might do better to call rain forests "the coral reefs of the land."

Life on the reef is so rich and varied that even the most fastidious scientists write with poetic abandon when describing this environment. In his 1930 lecture "Coral Reefs and Atolls," delivered at Boston's Lowell Institute, the Cambridge University biologist J. Stanley Gardiner opened his description of coral reefs by extolling "the beauty of their immense wreaths of green floating upon the sea, brilliant ultramarine in the hot sun, the whole dappled with the cloud shadows . . ."[20]

Several decades earlier, the great naturalist Alfred Russel Wallace, cofounder with Charles Darwin of the theory of evolution by natural selection, was traveling through the Indonesian archipelago collecting beetles, birds, and butterflies for sale and study. Wallace was taking a boat to his new base camp on the island of Ambon in the eastern part of the chain when he looked over the side and saw in the water "one of the most astonishing and beautiful sights I have ever beheld."

> The bottom was absolutely hidden by a continuous series of corals, sponges, actiniæ [sea anemones] and other marine productions, of magnificent dimensions, varied forms, and brilliant colours. . . . In and out among [the rocks and living corals] moved numbers of blue and red and yellow fishes, spotted and banded and striped in the most striking manner, while great orange or rosy transparent medusæ [jellyfish] floated along near the surface. It was a sight to gaze at for hours, and no description can do justice to its surpassing beauty and interest. For

> once, the reality exceeded the most glowing accounts I had ever read of the wonders of a coral sea.[21]

Notice Wallace's slight dig at others who had been moved to write with passion about the beauty of reefs, with the implication that what Wallace saw in Ambon *uniquely* justified his own effusion. This, too, is typical of scientists after a close encounter with a coral reef. They find the rhetoric of others a bit, well, overstated, and they imply that if their own writing departs from the coolly analytical prose demanded of science, it is only because reality compels them.

Darwin himself set this pattern a couple of decades before Wallace. After complaining about the "rather exuberant language" of "naturalists who have described, in well-known words, the submarine grottoes decked with a thousand beauties,"[22] the great scientist proceeds to rhapsodize on the miraculous nature of the coral atoll that arose before him in the Indian Ocean: "We feel surprise when travellers tell us of the vast dimensions of the Pyramids and other great ruins, but how utterly insignificant are the greatest of these, when compared to these mountains of stone accumulated by the agency of various minute and tender animals!"[23]

Personally, I'm willing to excuse nearly all impassioned writing when it comes to reefs—and not merely so that I can pursue my own florid prose. The profusion and spectacular diversity of life found on coral reefs simply elicits this reaction in humans. The scientist Gardiner was probably right when he said, "There is something in the psychology of mankind to which coral reefs never fail to appeal."[24]

Like love, the coral reef is a great mystery that sweeps over us, bypassing our rational minds entirely and eliciting feelings we didn't know were in us. The experience can be overwhelming. The head of an environmental organization devoted to preserving Florida's reef once told me that the first time he dived in a "forest" of elkhorn coral he was so moved that he hyperventilated and almost drowned.

I know the feeling. I was snorkeling on the reef that surrounds Australia's Heron Island, just one of the nearly three thousand coral reefs that make up the fourteen-hundred-mile-long chain known, collectively, as the Great Barrier Reef. I made my way out toward the reef crest an hour before sunset, when daylight creatures were still active. (The reef is like a factory, with day and night shifts, separated by a thirty-minute "quiet period" when few fish of either shift are seen.[25]) On the lagoon bottom, the sheltered and shallow reach between the island and the reef crest, there were innumerable black sea cucumbers, each up to a yard long, feeling their way slowly across the sand and leaving behind elongated fecal pellets of pure sand—purer than the sand that had entered their mouths, because the creatures remove most organic matter. Soon the sandy bottom gave way to a rockier environment. I drifted over a New Caledonian sea star—more commonly, but mistakenly, called a starfish—with an intricate mosaic of orange and yellow, the lines so finely drawn that it resembled cloisonné jewelry. I stopped to gently pick up a bright blue sea star, its long arms draped over algae-encrusted rock. I had expected it to go limp in my hands, but the creature was more rigid than I had expected; its five appendages retained the shape of the rock. I turned the sea star over in time to see its dozens of tubed feet retract along a groove that radiated down each arm. I returned the animal to its perch and swam on, passing over yet another, but very different, kind of sea star—a cushion star. Its mosaic pattern was brown and green, but the creature lacked apparent arms, looking like what its name suggests, a five-sided cushion. Cone shells, beautifully patterned but venomous snails capable of firing a poisoned tooth from one end of their shell, crawled over the hard bottom in search of a meal, probably worms or perhaps small fish. A honeycomb eel poked its black and white head out from its rocky den to see who was coming by, and then slowly withdrew.

As I continued swimming in a line perpendicular to the beach, a juvenile green sea turtle appeared from the right, spotted me, and veered off gracefully in the opposite direction, looking more like a bird than the reptile it is (but birds descended from reptiles, so the impression was apt). I must have approached the reef crest, for all at once living corals were everywhere, in the amazing variety of color and form that Wallace had described: leatherlike greenish corals, the colony folded in on itself many times; large tabletops of lacy coral structures, comprising perhaps a million individuals, their tiny tentacles still withdrawn and waiting for darkness; blue-tipped corals, and others with a reddish tint; and, strewn around the bottom, like upside-down mushrooms, *Fungia* corals.

Fishes were everywhere, too. An orange-and-white striped clownfish peered out from a tangle of purple-tipped stinging tentacles belonging to a sea anemone. Curiosity got the better of him, and as long as I remained immobile, the clownfish inched his way toward me. As soon as I moved, however, he tore back into the safety of the tentacles, like a base runner tagging up. A black-tipped reef shark swam lazily by, the prominent bulge of her abdomen identifying her as a pregnant female. After moving out of range the shark returned, eyed me for a moment, and disappeared again. Multicolored wrasses poked in and out of holes. Angelfishes of many varieties nibbled bits of algae from rocks and coral. Large, gaudily colored parrotfishes did more than nibble at the coral. They bit off entire chunks of coral, hard skeleton and all, using their fused, beaklike front teeth and grinding up the mixture of limestone, algae, and soft coral tissue with a specialized rock-hard device called a pharyngeal mill, located at the back of the throat. (One coral scientist has described the parrotfish's unique ability this way: "They'd eat a McDonald's parking lot to get the grease out of it."[26])

My favorite among the fish were the many species of boxfish, probably because their sharp-edged rectangular shape, the result of fused skeletal armor plates, seemed so unlikely. And perhaps I liked these fishes, among the most highly evolved in the sea, because they seemed as curious about my form as I was about theirs, allowing me to approach quite closely, sometimes initiating the contact themselves.

Intent on following a particularly colorful boxfish, I nearly swam right into a huge round table of coral—*Acropora valenciennesi*, a spectacularly well-branched variety that, from a few feet away, looks like the world's largest rack of elk antlers. The colony was round and dotted with dozens of species of multicolored fishes looping in and out of the white-tipped "antlers." I swam above, and hovered there, mesmerized. I lost track of time. With my arms and legs fully extended, my toes and fingers just barely reached the perimeter described by the circular colony. I felt like the Universal Man in the famous Leonardo drawing, the sketch of a man with legs and arms similarly outstretched and conforming to a perfect circle.

I floated there at the reef's edge. It was now the cusp of day and night. Below me was a perfect living circle of slender-branched corals, their tiny, gelatinous tentacles just now beginning to emerge for their nighttime feeding from a colony that was precisely my size. Above, if I turned my head, I could see a few faint stars venturing out, like the coral polyps below, into the tropical twilight. And rather than feeling alien in this exotic world, I was filled with the opposite sensation: I felt completely, if inexplicably, at home, as if I belonged there as much as the fishes or the sea cucumbers or the corals. It was as if all those years on land had been the sojourn in foreign territory and now, on the reef, I had arrived back home.

What lay behind those feelings, so irrational but at the same time so strong? Was it a genetic memory—the response of an animal whose ancestors, the earliest amphibians, had first lumbered out of these waters some 370 million years ago? Was it an emotional response to the overwhelming beauty of the place? Perhaps the exertion of swimming a kilometer had stimulated my pituitary gland to release an abundance of endorphins, and these feelings were merely a side effect of a brain drunk on neurotransmitters. I don't know. All I know was that in that moment I didn't want to return to land, ever.

But of course I did return. I swam slowly and regretfully back to shore, cursing my lack of gills all the way. I passed two huge octopuses, mottled red and white, who, as I approached, blanched completely white and reared up, their heads as big as basketballs. These shy creatures, mollusks without shells, are able to change their skin color by contracting or expanding specialized cells called chromatophores, each one containing minute sacs of dye. The ghostly twins held their ground, literally, their sixteen tentacles (eight apiece) sliding nervously through holes in the rock as they directed large and humanlike eyes at me. I moved off slowly. I was so excited about this excursion that when I had almost reached the beach I did a spontaneous somersault in the water—

nearly landing on two huge rays resting on the sandy bottom. Climbing at last onto the beach, I lay there for a long time, exhausted and exhilarated and still under the reef's spell, looking up into a sky slowly being set ablaze with stars.

❄

There is a word to describe what Gardiner, Wallace, Darwin, and the rest of us have felt when in the presence of the reef: *awe.* Confronted with the reef, awe is the most appropriate response. It is probably in our nature.

It is also, apparently, in our nature to destroy that which we hold in awe.

Perhaps we want to subdue whatever arouses those powerful feelings of awe, to master it, so that we are no longer threatened by such primal emotions. I'll leave that one to psychologists and philosophers. Regardless of the motivation, however, we are doing a rather good job of destroying coral reefs around the world. According to recent estimates, "about 10 percent [of the world's coral reefs] have already been degraded beyond recovery and another 30 percent are likely to decline significantly within the next 20 years."[27] In some areas human activity has destroyed entire reefs, converting them into algae-covered rubble. Who knows what species, known and unknown alike, have already been wiped out? Who can say which ones will be winking out in the near future, their intricate genetic codes, formed over millennia, suddenly terminated, their potential for medical and other human uses forever gone, their roles, perhaps key to the complex patterns within the reef community, ended? Like so much else about the coral reefs, we don't know the answers to these questions. But we had better start looking for them, for a lot is riding on those answers.

The reef and our bond to it makes a fine metaphor for our larger relationship with nature. But there is more going on here than metaphors. For a variety of reasons, which I will attempt to explain throughout this book, our fate is bound up with that of the reef.

Of course, if we are to save the reefs, we must understand them better. But here we have come full circle. We have returned to our initial question, the one that faced our computer-reliant ecologist: How do you comprehend something as complex as the coral reef?

Probably the only way is to start at the beginning. You contemplate the simplest, the most fundamental element of the coral reef. You begin with the individual coral polyp and move outward from there.

2

Animal, Mineral, Vegetable

There is nothing that God hath established in a constant course of nature, and which therefore is done every day, but would seem a Miracle, and exercise our admiration, if it were done but once.

—John Donne

The lessons we learn earliest we learn the best. They are, consequently, the hardest to shake. For example, I still divide the world as I did when I was a child, into three fundamental groups: animal, mineral, vegetable. Having learned the rules of scientific classification has done nothing to change this quirk. Animal, mineral, vegetable: this was the trifocal lens through which the undifferentiated world first swam into focus, and it remains with me still.

Scientists who classify organisms, called taxonomists, have developed a far more intricate system, of course. The basic version, in descending order from the most general to the most specific, is this: kingdom, phylum, class, order, family, genus, species. There is also a more complex rendition of this hierarchy that contains sub- and super- versions of each category—as well as an "infraclass," a "cohort," and even the rarely used "tribe." After drilling myself on these proper divisions, it gave me an intellectual *frisson* to discover that Linnaeus, the eighteenth-century Swede who is generally recognized as the father of modern taxonomy, divided the world into three basic kingdoms: animal, mineral, and vegetable.

From this starting point, Linnaeus and his descendants have subdivided and reorganized this "family tree" ever more finely. Stony corals, the kind that build reefs, presented a unique problem. Though corals were long considered marine plants, two sixth-century observers noted that they bore certain similarities to animals. They decided to hedge their bets, calling corals zoophytes, literally "animal-plants."[1] More than a millennium passed before a French naturalist named Peyssonnel published a paper declaring the coral to be an animal, a category in which it still resides.

And yet this solution seems a bit artificial. At a gross intuitive level, corals are a pretty miserable excuse for an animal. What we tend to think of as an individual coral, a large, branchlike structure or a boulder, is really a colony of tens of thousands of tiny animals, most of them just a few millimeters across. Those small animals are extremely simple affairs, anatomically speaking. They are composed of just two layers of true cells, where nearly all other multicellular animals have at least three. A coral polyp, the proper name for an individual coral, is basically a tiny ring of gelatinous tentacles fluttering above an equally small, internally rippled sac. Hard corals also have a skeleton, or corallite, at their base, into which the polyp retreats during the day. Corals lack nearly all organs and structures generally associated with more familiar animals such as dogs or squirrels or, for that matter, people. Corals don't have legs or ears or eyes or even what most of us would recognize as a true nervous system. Almost without exception, they cannot move from one place to another (at least not as adults under normal circumstances and not under their own power). But scientists will argue quite properly that there are a great many organisms grouped under the kingdom Animalia that do not remotely resemble dogs or squirrels or us. What binds us animals together, they will say, is primarily a shared characteristic. A deficiency, really: we are all unable to convert sunlight into food, to photosynthesize as plants do. We are not "primary producers"—hence, we need to eat.

And corals—like dogs, squirrels, and humans—do eat other beings. At night their tentacles emerge to shoot paralyzing darts into

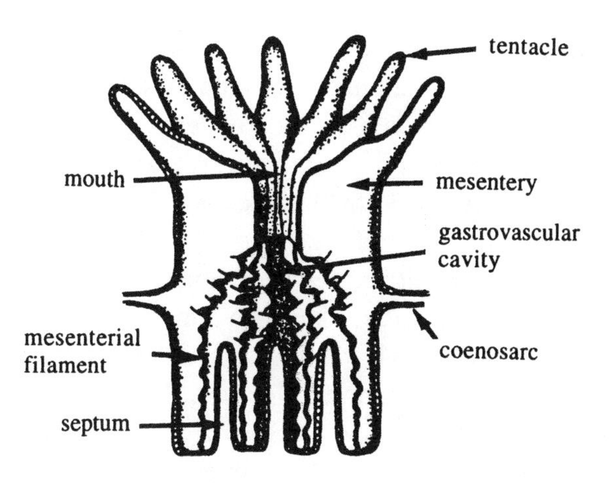

Figure 1. Polyp Cross Section

their prey, mostly zooplankton (microscopic animals carried along by ocean currents), pulling these morsels into their primitive mouths, an opening that leads into an equally primitive stomach where specialized cells secrete digestive enzymes. If this were the only way in which corals obtained nutrition, then their presence among the rest of us animals might not seem quite so strained. A distant relative to be sure, but a relative nonetheless.

But reef-building corals do *not* rely upon eating other beings. In fact, only a fraction of their nutrition, somewhere in the neighborhood of 2 percent,[2] comes from consuming other organisms. This question of coral nutrition had scientists stumped for quite some time, and continues to spark a certain amount of controversy. Researchers measuring corals' nutrient intake against their growth and energy expenditure saw that something was clearly wrong with the resulting picture: more energy was coming out of the system—the corals—than was going in. And yet the corals thrived, oblivious of their starvation diets. It was just the first of many tricks corals had up their gelatinous sleeves.

Scientists couldn't find their food source because they were looking in all the wrong places.

They searched the environment surrounding the coral, a logical place to look when considering an animal. But the primary food source of hard corals lies within them. Inside each coral polyp lives a microscopic form of algae called zooxanthellae. Coral scientists have affectionately nicknamed them "zoox" (rhyming with "folks"). There has been much confusion about the lineage of zoox. It is now generally believed that zoox exist in several different genera and even in several families, perhaps in different classes as well. What this means is that the decision to include an algae in the group zoox isn't based on its evolutionary heritage, but on the algae's *behavior.* Some scientists prefer to use the adjective *zooxanthellate* algae, steering clear of the noun *zoox* altogether. Non-purists (including the author) use the short form, but with the understanding that it describes behavior, not necessarily ancestry.

Like other plants, zoox photosynthesize organic compounds from sunlight. Then they pass along the bulk of their food (up to 98 percent of the total[3]) to their coral hosts—"paying rent" one coral physiologist calls it. Without the nutrients supplied by the zoox, the corals would die. Scientists have conducted numerous experiments to test this hypothesis. Corals were covered so that no light could reach them. After hanging on for a few days or even weeks, the corals succumbed. A similar situation occurs in nature when zoox abandon their corals (a little-understood process called *bleaching,* which occurs under certain stressful situations). The result is the same: the corals starve to death.

Since nearly all coral tissue is transparent, it is the zoox that give hard corals their color. Under a microscope, coral polyps appear bloated with the golden brown zoox—with as many as a million individuals packed into a single square centimeter of coral.[4]

This is, of course, a two-way street. The zoox don't stick around donating food because they are generous. We know what the coral gets: food in the form of sugar lipids and oxygen. But what about the zoox? What do they get? Their own form of food: carbon dioxide (CO_2) and nitrogen, in the form of ammonia, both waste products from the coral's metabolism. The zoox need these just as much as the corals need the food that the zoox supply. Corals, with their simple tissue and their stony skeletons, also provide zoox a safe haven from the many herbivorous predators lurking on the reef.

Biologists call this kind of long-term relationship between organisms *symbiosis.* Think of it as a kind of marriage. Not all marriages are made in heaven, and neither are all symbiotic relationships. In a given marriage, one individual may reap all the benefits—while the other one just suffers through. Biologists have a term for this kind of arrangement: *parasitic symbiosis.*

When both human partners benefit from the relationship, we call it a "good marriage." In nature, biologists describe this relationship as *mutualistic symbiosis* or simply *mutualism.* There are many fascinating examples of mutualism in nature. One well-studied example is the relationship between the bullshorn acacia tree and a species of ant that lives on it.[5] The acacia provides nectar to the ant colony, and the ants swarm to attack any herbivores that threaten "their" tree. Both ant and tree benefit from the relationship.

There are many other examples of mutualism in nature, but none more intimate than the one that exists between corals and zooxanthellae. Most mutualistic relationships occur between clearly discrete organisms, such as the acacia tree and the ant. But in the case of the coral and the zoox, the symbionts are virtually inseparable, coexisting in a single unit, the coral polyp, from generation to generation, evolving together over thousands of years. Some "baby" corals, larvae called planulae, carry zoox within them as they leave the colony, a housewarming present from their parents.[6]

The sixth-century observers who classified corals as zoophytes were on to something—although perhaps for the wrong reasons. It is fair to ask, then, if corals and zoox are separate entities. After all, some coral species contain so many of the algae that, when weighed, the biomass of the zoox equals that of their hosts. In the strictly scientific sense, certainly they are distinct organisms. And yet they are so tightly interwoven that in the real world of the coral reef they make no sense apart.

Zoox perform one more vital service to corals: they speed up the process by which corals build their stony skeleton, putting down layer after layer of calcium carbonate, a form of limestone. It is this process of calcification that physically constructs coral reefs, the literal bedrock of the coral reef. The process occurs on such a huge scale that coral reefs are responsible for half of all calcium sedimentation occurring in the ocean.[7] Still, there are many agents constantly wearing away at reefs. Worms and sponges bore through them. Parrotfish crunch at the hard rock to get at the algae and coral tissues. Waves relentlessly pound the reef into sand. Zoox make the process of coral calcification take place two to three times faster than it would otherwise occur. If not for the zoox's help, the rate of erosion would overtake the rate of construction and the limestone reefs would simply disappear, at least in many places. Certainly this is another reason to sing the praises of zooxanthellae, but this new element suggests something even more important.

The "animal-plant" called coral is more properly an "animal-plant-mineral," with each of the three playing a critical role in the survival of the others. If the name *zoophyte* made some sense, then *zoophytelite* (animal-plant-stone) makes even more, for nowhere in nature are the three basic elements of the planet woven together in such a close fashion—nor with such spectacular results.

Darwin was right when he observed that the pyramids are "utterly insignificant" when compared to the underwater mountains of limestone that coral and their zooxanthellae have erected, secreting one thin layer of calcium carbonate at a time, at an average of a fraction of an inch each year.[8] Yet all those fractions add up over time, and there are places on Earth in which corals live on structures built by their ancestors and predecessors, going down 7,220 feet, well over a mile.[9]

The coral polyp—this diminutive and deceptively simple creature, this enchanted braid of animal, mineral, and vegetable—is responsible for the largest biogenic (made by living organisms) formation on the planet[10] and the most complex ecosystem in the sea: the coral reef.

Living coral reefs are found in the clear, warm, mostly tropical waters of more than one hundred nations throughout the world.[11] Although there is still much more to learn about the subject, some of the environmental requirements for coral reef growth have been known in a general way for some time. For example, between 1838 and 1842, while circumnavigating the globe on the U.S. Exploring Expedition, an American geologist named James Dana noted that "the temperature limiting the distribution of corals in the ocean is not far from 66°F."[12]

According to Dana, this explains why coral reefs are found in the Bermudas, which, at 33° latitude, is far north of their coral's "normal" range, and not in the Galápagos Islands, which are well within the tropics. Bermuda lies in the warm currents of the Gulf Stream, whereas the Galápagos are washed by the cold currents of coastal South America.

Dana was essentially correct—although he was wrong about the Galápagos: corals *do* grow there. Look at a map of the earth showing the places in which the average surface temperature for seawater is at least 68°F (see figure 2). Notice how most coral reefs fall within the boundaries of this 68°F isotherm. Yet the actual role played by temperature is open to debate. It's clear that reefs don't thrive in waters that are too cold (or too warm). But sunlight itself, specifically the amount of solar radiation striking the ocean, not temperature, is almost certainly the more important element limiting the growth of coral reefs.[13]

Reefs also need clear water, free from sediments that prevent sunlight from reaching corals and their photosynthesizing algae. Because most sediments come from land, born by rivers emptying into the sea, thriving coral reefs are generally found far from such influences. Just as important is water free from nutrients—what is

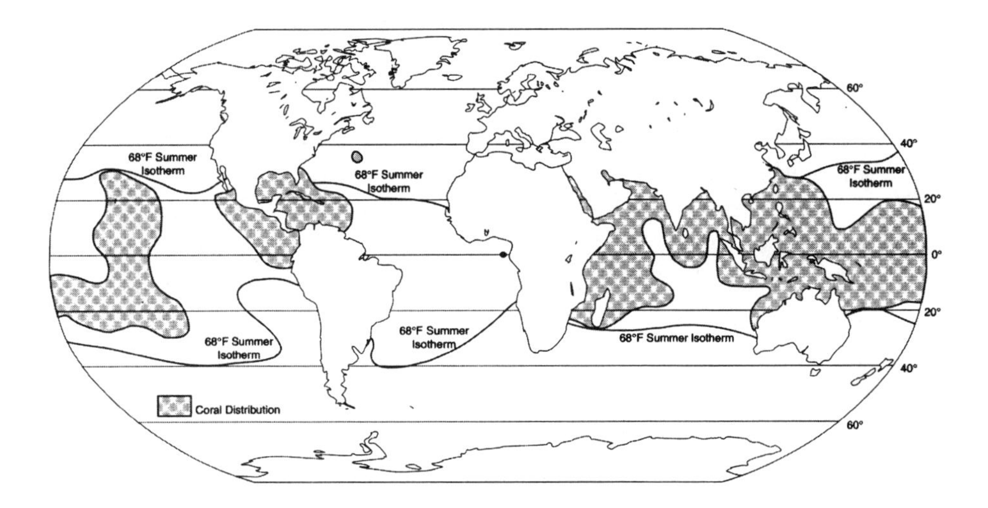

Figure 2. Coral Distribution

known as an oligotrophic environment. Eutrophic waters (those containing high levels of nutrients), by contrast, provide a movable feast for the tiny drifting organisms called phytoplankton, a scenario that is devastating for corals, which are easily overgrown by these phytoplankton "blooms."

These characateristics, and a suitable hard bottom in shallow waters, are the primary environmental needs for coral reefs, and they are found to varying degrees throughout the world. Bathed by the warm waters of the Gulf Stream in North America (as Dana pointed out), coral reefs form a large living arc just off the Florida Keys, and surround the islands of the Bahamas. In the Caribbean they form the second largest barrier reef in the world (off the coast of Belize), and border countless islands, large and small, from the Caymans in the west to Cuba in the center, to Dominica and Martinique in the east, and off the coasts of the bordering countries of Mexico, Honduras, Nicaragua, Costa Rica, Panama, Colombia, and Venezuela.

In the Mideast they thrive all along the desert coastlines of the Red Sea and the Arabian Gulf. In the Indian Ocean, coral reefs stretch down the shores of eastern Africa—the coasts of Somalia, Kenya, Tanzania, and Mozambique—across to the islands of Madagascar, the Seychelles, and the Maldives, and over to Sri Lanka and India.

Coral reefs are sprinkled across the wide Pacific Ocean, in the waters surrounding French Polynesia, American Samoa, Fiji, and Hawaii; throughout Micronesia and Melanesia; from Japan in the north down to the immense Great Barrier Reef system of Australia in the south.

In the Indo-Pacific region, in the nations of Southeast Asia (including Indonesia, the Philippines, Malaysia, Singapore, and Thailand), coral reefs reach their greatest abundance. It is estimated that this one area is home to nearly one-third of the world's coral reefs.[14]

Estimates of the total area covered by the coral reefs around the world vary widely, from a conservative 88,800 square miles to a high-end figure of 580,000 square miles. Predictably, the estimate most widely cited is somewhere in the middle: about 240,000 square miles.[15] (These disparate figures have more to do with how a coral reef is defined than with any real arguments about where they exist.)

Regardless of which number is used, the importance of coral reefs cannot be measured by square miles alone. After all, we're talking about an area the size of which is somewhere between Minnesota and Alaska, with the most accepted estimate putting it in the same league with Texas. Not insignificant, certainly—but not huge either. And that's comparing it to land area, a concession to us terrestrial types. Compared to the vastness of the oceans, coral reefs are mere specks. But they are specks in the same way that the Earth itself is a speck in our solar system: unique, luminous, alive.

Their importance to humans is immense. Hundreds of millions of people around the globe depend upon coral reefs for food, for livelihood, for protection from battering waves that would erode entire islands into oblivion, and even for providing social stability in traditional cultures.[16]

In the Florida Keys, the living reef adds hundreds of millions of dollars annually to the local economy.[17] In the Philippines, between 10 and 15 percent of the animal protein in people's diets comes from the reef.[18] For the residents of many islands, coral reef fishes provide 100 percent of their animal protein.[19]

With perhaps millions of unexamined species living on and within them, coral reefs are medicinal treasure troves, similar to, and perhaps greater than, tropical rain forests. This isn't simply speculation. Corals are already used as a bone-graft substitute[20] and as a natural source for prostaglandin, a lipid with a variety of medical uses.[21] Other reef creatures have been found to contain compounds that fight cancerous tumors, leukemia, and a multitude of other diseases.[22]

Only recently, scientists discovered that coral reefs provide another important resource, as historical climatic recorders. In the same way that tree rings provide a record of weather conditions, polyps keep a record, etched literally in (lime)stone, of seawater temperature, rainfall, and other environmental conditions. These "coral thermometers" are remarkably accurate—down to a monthly resolution—going back tens of thousands of years. Because of this, says one climatologist, "corals hold the Rosetta Stone to understanding climate change,"[23] past and future.

Scientists from other disciplines are just as enthusiastic about the role that coral reefs can play in basic biological research. One ecologist extolled their role as "marvelous natural laboratories for exploring questions of fundamental interest in evolution, ecology and animal behavior."[24]

But forget about humans for a moment. Coral reefs are critically important to the planet, with or without us. They shape the shorelines of islands and continents. And even though they constitute a minute portion of the global ocean, reefs exert a tremendous influence on the entire marine environment. The very chemistry of the sea is altered by corals' ability to fix calcium. Reefs provide a habitat for enormous numbers of marine species, and act as spawning grounds for perhaps tens of thousands more, species that spend the rest of their lives in the open waters of the ocean. What would the ocean be like without them? It's impossible to say, for coral reefs are the very soul of the sea.

3

Darwin in Paradise

> Every one must be struck with astonishment, when he first beholds one of these vast rings of coral-rock, often many leagues in diameter, here and there surmounted by a low verdant island with dazzling white shores, bathed on the outside by the foaming breakers of the ocean, and on the inside surrounding a calm expanse of water, which, from reflection, is of a bright but pale green color.
>
> —Charles Darwin

Many scientists had studied corals, but there existed no grand theories on reef formation as a whole until a twenty-six-year-old gentleman-naturalist named Charles Darwin had his thoughts on the subject presented at a meeting of the Geological Society of London on the last day of May 1837. He published his full work on the topic five years later in a book titled *The Structure and Distribution of Coral Reefs.* It was a significant milestone for the young man—his first scientific book containing his first original theory—and it caused the initial stir in a career that was to be filled with roiling debates. He had just returned from a life-changing voyage, a five-year circumnavigation of the globe aboard the HMS *Beagle.*

Looking back from the distance of old age, Darwin would write of this book: "No other work of mine was begun in so deductive a spirit as this, for the whole theory was thought out on the west coast of South America, before I had seen a true coral reef."[1]

Let me put this as kindly as possible: Darwin often treated the truth as if it were elastic. It is likely that his statement above is one example of that lifelong tendency. Not the "whole theory," but likely the basis of it came to him bit by bit as he explored the mountains of Chile and Peru. What he saw later in the voyage helped him refine his theory, first in Tahiti and then in the Indian Ocean at the tiny islands of Cocos-Keeling and, even later, off the African mainland, on the island of Mauritius.

Regardless of the process involved in arriving at his theory, the work represents the extraordinary beginning of a brilliant career. It was also a natural for Darwin, combining his new and growing love for geology with his boyhood interest in biology. Little "Bobby" Darwin had spent his childhood avidly collecting and identifying beetles and bird's eggs from the garden and woods surrounding his family's estate outside of Shrewsbury, England.[2]

His interest turned toward marine zoology, and to corals in particular, while he was a second-year medical student attending Edinburgh University. There he fell under the influence of an instructor named Robert Grant, a rather stiff and dry lecturer, who came to life only while studying the simple sea creatures growing on the rocks off the rough Scottish coast. Grant communicated this zeal to the younger student. Grant was the leading authority of his day on sponges, giving them their family name, Porifera.[3] He also studied corals. In addition, the Edinburgh professor was a proponent of a form of evolution, and saw corals (or zoophytes, as they were still called by many) as the missing link between plants and animals.[4]

This marked the beginning of an interest that would stay with Darwin for many years. At one point he even considered focusing exclusively on the minute marine animals. As he sat writing at his desk in his cramped quarters aboard the *Beagle*, sailing along the coast of Chile, he informed his sister Catherine that "I have lately determined to work chiefly amongst the Zoophites or coralls [*sic*]. It is an enormous branch of the organized world; very little known or arranged & abounding with most curious, yet simple, forms of structures."[5] And this was still long before Darwin had studied living corals in their natural element.

For the first several months of the voyage, Darwin sailed south through the Atlantic, working out a relationship with his shipmates (especially with the temperamental Captain FitzRoy), and simultaneously battling homesickness and seasickness—two illnesses he would not conquer during the journey. There were a few stops at islands along the way, and Darwin made full use of his time ashore, but it was not until the *Beagle* reached the continent of South America, anchoring in Brazil in February 1832, that the young Englishman (he had just turned twenty-three) was able to begin roaming the strange lands he had dreamed of as a child wandering around Shrewsbury.

"Delight itself," he wrote, "is a weak term to express the feelings of a naturalist who, for the first time, has wandered by himself in a Brazilian [rain] forest."[6]

As the voyage continued down the eastern coast of South America, Darwin collected and preserved specimens of beetles, butterflies, rocks, fish, and birds. He observed with feverish intensity the landforms as they slipped by, and theorized about how they came to be there. He rode with gauchos, the cowboys of the Argentinian pampas, collecting fossils along the way, and spent Christmas waiting out a howling storm in Patagonia, at the continent's southern tip.

Then began what was to be one of the most important parts of the entire voyage, the slow journey up the west coast, the Pacific margin of South America. This certainly would be *the* most significant part of the trip, as far as Darwin's theory on coral reefs is concerned.

The *Beagle* anchored in the Bay of Valparaiso, in central Chile, in the dark of night on July 23, 1834. When he emerged from his small cabin the next morning, Darwin was greeted by the spectacular sight of bright blue skies and sunshine cascading upon the red tile rooftops of the city and upon the snow-capped volcano Aconcagua, which rose magnificently in the distance. At nearly 23,000 feet, it is the highest peak not just in the Andes, but in the western hemisphere.

The Andes beckoned to Darwin as nothing else had, not even the rain forests of Brazil. The mountain chain, which he had glimpsed on the eastern side of South America, ran the length of the continent, from Panama in the north to Tierra del Fuego in the south, a distance of 4,500 miles.

On August 14 he left Valparaiso and headed off on horseback into the Andes, accompanied only by a crew member Darwin hired as an assistant. The trip, which lasted slightly more than a month, would change his life. And not all for the good. Darwin developed an illness, the source of which was never firmly established, that would affect his health until he died, forty-eight years later. It was the first serious illness of his life, and for weeks he lay in bed in Santiago, feverish and in extreme pain.

But the trip was also an intellectual revelation to Darwin. It was during this twisting journey, through high mountain passes distant from the ocean, that Darwin began to develop his theory about the formation of coral reefs. His very first night out of Valparaiso, he saw towering beds of fossilized seashells, on land far from the sea.

"[A]t the height of a few hundred feet old-looking shells are numerous," he wrote, "and I found some at 1300 feet."[7]

Many of the fossils were stuck in what Darwin at first thought was a mass of decomposing leaves. However, "I was much surprised to find under the microscope that this vegetable mould is really marine mud, full of minute particles of organic bodies."[8]

How did sea creatures arrive on a mountainside? To answer this, Darwin turned to the theories of Charles Lyell, an eminent, though controversial, geologist. The younger man carried Lyell's book *Principles of Geology* with him on the long voyage, a gift from Captain FitzRoy. Lyell's influence on Darwin's thinking was immense. As Darwin's biographer Janet Browne succinctly puts it: "Without Lyell there would have been no Darwin."[9]

Lyell postulated that the earth was not a static lump of rock, but a dynamic, ever-changing, finely balanced explosion of forces. This dynamism was seen in the slow but powerful uplift of land in some parts of the world, and, to counterbalance the swelling, the equally slow but dramatic subsidence of land in another part of the world. It was a revolutionary way of seeing the world, one viewed by many (although, oddly, not by Lyell himself[10]) as a challenge to scriptural literalism, which saw the world as an unchanging creation of God.

Darwin was immediately and completely captivated by this idea.

The fossil seashells and marine detritus high in the Andes of central Chile closed the case as far as Darwin was concerned: "The proofs of the elevation of this whole line of coast are unequivocal."[11]

Lyell had provided an essential key to the puzzle of coral reefs. Over a year later, on November 15, 1835, another piece fell into place. The *Beagle*, having completed the voyage up the Pacific coast of South America, surveying the Galápagos Islands far off the coast of Ecuador, sailing halfway across the Pacific and south toward Australia, finally dropped anchor at Tahiti—"an island which must for ever remain classical to the voyager in the South Sea," recorded Darwin.[12]

Two days later, after a difficult climb partway up one of the two volcanic cones that dominate Tahiti, Darwin gazed west toward the neighboring island of Moorea. He saw a scene of incomparable beauty—which posed multiple and bedeviling questions.

"The island," he wrote, "with the exception of one small gateway, is completely encircled by a reef. At this distance, a narrow but well-defined brilliantly white line was alone visible, where the waves first encountered the wall of coral. The mountains rose abruptly out of the glassy expanse of the lagoon, included within this narrow white line, outside of which the heaving waters of the ocean were dark-colored."[13]

What could account for this pattern, which Darwin saw repeated (at a distance) several times on the voyage: a living ring of coral separated by a shallow lagoon from a soaring mountain at the center? Even more curious were the coral atolls, as natives called these islands, entirely lacking a central mountain or significant land of any kind; just a ring of coral reefs, and perhaps a narrow, sandy shoreline barely above the water, surrounding a shallow lagoon, the whole encompassing many square miles. Lyell had earlier hypothesized that corals grew along the rims of submerged volcanos, but this theory could not adequately explain what Darwin saw. As the young man later wrote his sister, "The idea of a lagoon island, 30 miles in diameter being based on a submarine crater of equal dimensions, has always appeared to me a monstrous hypothesis."[14]

Darwin didn't have a chance to investigate the reefs of Tahiti or Moorea. The *Beagle* had to continue on to New Zealand and westward to Australia. In fact, he had yet to scrutinize, up close, a single coral reef in a journey that had already covered many thousands of miles. This lack was supremely frustrating to the young naturalist who was considering making a career out of the study of corals.

Finally, on April 1, 1836, he had his chance. The *Beagle* arrived at the Cocos-Keeling Islands—a group of twenty-five islands, nearly six hundred miles southwest of Java in the Indian Ocean. Darwin spent his first few days investigating and admiring the local flora and fauna: ". . . [N]othing could be more elegant than the manner in which the young and full-grown cocoa-nut trees, without destroying each other's symmetry, were mingled into one wood. A beach of glittering white sand formed a border to these fairy spots."[15]

Then, on April 4, Darwin began his first field trip to a coral reef, not an easy task in those days, before scuba or even snorkeling equipment. He must have presented an amusing spectacle to the people of the islands: a properly dressed Englishman using a "leaping-pole" (Darwin's phrase) to hop between coral formations out to the reef's crest.[16] For the next several days, Darwin roamed the coral reefs and lagoons of the area. It is clear from his travel diaries that he was having the time of his life. He writes in giddy astonishment of giant sea turtles swimming in the crystal-clear lagoon. He describes the multicolored parrotfish feeding on the corals and the giant clams "into which, if a man were to put his hand, he would not, as long as the animal lived, be able to withdraw it."[17]

But most of all he admired the work of the coral polyps themselves.

> It is impossible to behold these waves without feeling a conviction that an island, though built of the hardest rock . . . would ultimately yield and be demolished by such an irresistible power.

> Yet these low, insignificant coral-islets stand and are victorious: for here another power, as an antagonist, takes part in the contest. The organic forces separate the atoms of carbonate of lime, one by one, from the foaming breakers, and unite them into a symmetrical structure. Let the hurricane tear up its thousand huge fragments; yet what will that tell against the accumulated labor of myriads of architects at work night and day, month after month? Thus do we see the soft and gelatinous body of a polypus, through the agency of the vital laws, conquering the great mechanical power of the waves of an ocean . . .[18]

The coral reef, visited at last, was everything Darwin had hoped it would be: beautiful beyond words, crawling and swimming with a multitude of creatures, large and small, and all of it dependent upon the miraculous biochemical processes of tiny organisms. Clearly, what most intrigued Darwin was the ability of corals to alter, and in fact to create, the physical landscape.

They arrived back in England on October 2, 1836, nearly five years after setting sail. Darwin dived into his work, sorting fossils sent on ahead from his travels, revising his diaries to form a narrative of the journey aboard the *Beagle*, and first a paper and then a book on coral reefs.

His theory was as ambitious and sweeping as those of his mentor (and now friend), Lyell. Darwin was planting a flag out on the margins of scientific thought, one that could be clasped or knocked down, but not ignored. It was perfect training for the intellectual battles to come.

The young traveler sought to explain the formation not just of atolls, but of two additional categories of reefs characterized as *barrier* and *fringing* reefs.

His theory of reef formation, like the best of science, was elegantly simple. He hypothesized that the three types of reefs were really of one kind—only viewed at different stages of development. According to this way of thinking, to see a fringing reef as something wholly different from a barrier reef, and to say that a barrier reef is unrelated to an atoll is wrong—as silly as insisting that a baby, a teenager, and an old woman are three different animals.

Darwin's primary contribution was to introduce the idea that coral reefs evolved (to use a loaded word) over time.

First come fringing reefs, such as Darwin saw clinging fast to the shore of Mauritius, in the Indian Ocean off southern Africa. Here, in shallow waters, coral communities begin their life, growing upward toward the ocean surface and then horizontally out to sea. There may be a narrow channel of water separating the coral from the land, but nothing too large.

Darwin saw that this seemingly static picture belied tremendous, if unhurried, activity. The key is to view the Earth as though using time-lapse photography—and with extremely long pauses between exposures. As the land sinks ever so slowly into the ocean (à la Lyell), the corals continue growing toward the surface. The community survives only by matching the rate of subsidence with its own upward development. Darwin understood that when reef-building corals sank below a certain level (around 165 feet) they died, although he didn't know why, lacking knowledge of their symbiotic zoox. But he *did* know that their limestone skeletons became the foundation for the live veneer of corals thriving closer to the surface.

As the sloping land continued to sink, reasoned Darwin, the corals continued to grow upward. Because the sinking volcano was wider at the base than at the top, the corals were now at a greater distance from the shore. Voilà: a barrier reef. In Darwin's words, the corals resembled "a wall with a deep moat within."[19]

The last stage was the coral atoll. The volcano had disappeared below the waves entirely, and left behind only the corals, the minute polyps constantly secreting layers of calcium carbonate and staying near the surface. What land existed was the result of accumulating coral rubble and sand. Later, vegetation would take hold.

Even though Darwin's theory took Lyell's work in a different direction, it was based on the older man's labors, much as living coral grows from a foundation built by its predecessors. Lyell, not surprisingly, was impressed: "I am very full of Darwin's new theory of Coral Islands," he wrote a colleague soon after the first paper was delivered. "Let any mountain be submerged gradually, and a coral grow in the sea in which it is sinking, and there will be a ring of coral, and finally only a lagoon in the centre. . . . Coral islands are the last efforts of drowning continents to lift their heads above water."[20]

The proof or refutation of those grand theories would have to wait more than a century for technology to catch up with Darwin's thinking. According to Darwin's theory, far below the layers of limestone accreted by corals was basalt, rock of volcanic origin. Darwin was a wealthy man, but he didn't have the money necessary to test this hypothesis by deep drilling. A year before he died, the old naturalist wrote his friend (and frequent critic) Alexander Agassiz: "I wish some doubly rich millionaire would take it into his head to have borings made in some of the Pacific and Indian atolls, and bring home cores for slicing from a depth of 500 or 600 feet. . . ."[21]

The first attempt to drill through ancient coral down to basalt occurred soon after Darwin's death. The results did not bode well for his theory. The Royal Society (of London) made several drillings on Funafuti, a pear-shaped atoll in the western Pacific nation of Tuvalu, between 1896 and 1898.[22] The expedition obtained core samples down to a depth of 1,114 feet, nearly twice the depth Darwin had suggested, but never reached the basaltic layer he predicted.[23]

Between 1934 and 1936 a Japanese team drilled down through 1,416 feet on the island of Kita Daito Jima, east of Okinawa. They found only limestone.[24]

In 1952 the "doubly rich millionaire" Darwin had pegged his hopes on finally arrived in the unlikely person of Uncle Sam. As part of its nuclear weapons testing program, the United States Atomic Energy Commission ordered deep drilling of limestone reefs around Enewetak Atoll in the Marshall Islands, more than two thousand miles northwest of Funafuti. Five years earlier the Americans had drilled on the nearby Bikini Atoll (another of the Marshall Islands), down to 2,556 feet without striking anything other than limestone,[25] but the team at Enewetak was prepared to go even deeper.

Enewetak is a large, nearly circular atoll, comprising some forty tiny islands surrounding a lagoon that encompasses almost four hundred square miles.[26] In the summer of 1952 the Americans landed on the mere speck of an island named Elugelab, which rose a few feet above sea level on the northern tip of the atoll. The team and all their gear nearly overwhelmed the small island. The group was led by Harry Ladd, a palaeontologist with the U.S. Geological Survey. It included at least a half-dozen other scientists (including a geophysicist from Los Alamos Scientific Laboratory, the nuclear weapons R-and-D facility) as well as twelve men assigned to drilling crews and their supervisor.

The year before, the Americans had used a twenty-five-foot rig mounted on the back of a truck to drill on Bikini. The current project was far more ambitious, however, and required a far larger apparatus—too large to fit onto a truck. The "big rig" looked like an oil derrick, rising more than one hundred feet into the air and supporting a powerful drill tipped by a three-cone rock bit.[27] There was also a diamond bit to take core samples from different depths.[28] A series of pumps were hooked up to the drill to bathe the bit constantly with cooling water and flush out ground sediments. The entire structure was mounted on a large trailer and towed out to the drilling site, a bit more than one hundred feet from the edge of the lagoon and slightly less than that from the sea.[29] Drilling would be continuous, with three-man crews putting in eight-hour shifts, and a geologist remaining at the site at all times.

When all was ready, Ladd instructed the first crew to begin. The drill was turned on and slowly lowered to the coral surface, where it immediately began chewing up coral limestone and boring a hole, which was named, according to the poetic sensibilities of government scientists, *F-1*.

Science fiction writers have thought up many ingenious devices; one of the most memorable of these, and the one that lent its name to almost all such contrivances imagined thereafter, was H. G. Wells's creation: the Time Machine. The ability to travel through that least tangible of dimensions, time, has captivated writers and readers for more than a century. Despite its low-tech appearance—all the cables and guywires and diesel-operated pumps—that's what the drill set up at Elugelab was: a time machine.

Every inch the whirling bit descended into the coral rock was also a trip backward through time. This would have been especially evident to Harry Ladd, who examined the materials brought up by the drill and calculated when they would have been deposited.

No one recorded how long it took to drill through the first ten feet. They were probably too busy making sure that the apparatus was boring straight. But they did record the time it took for the drill to bore through the next ten feet: four minutes. In moments the drill had traveled back past the time of Darwin's death, past the announcement of his theory, past his sickness in the Andes, past his birth. In seconds, Darwin's entire life was traversed by the spinning metal cones. And the drilling had just begun.

By the time someone first thought to write down drilling rates, at the ten foot mark, the drill had already sliced through more than a thousand years of coral growth.

At two hundred feet the drill was boring through limestone that had been formed long before the peak of the last ice age.

After eight hours of continuous drilling, the three-man crew was changed. The men had drilled through approximately 850 feet—and traveled backward in time over 100,000 years. Not a bad day's work.

But there was still no sign of basalt, the form of volcanic rock most likely to be underlying the reef limestone—if Darwin's hypothesis were correct.

Ladd made meticulous notes about the materials found at each level. For example, in his log for the depth of 190–280 feet, he entered: "Coral-head limestone; fragments of massive corals (*Porites*); finer material consists of small mollusks, Foraminifera, and segments of *Halimeda*; some yellow calcite; mollusk molds rare."[30] All of these items are common to a reef environment.

As you scan down Ladd's log for F-1 today, what is most striking is the uniformity of the material found as the drill went ever deeper. The proportion of mollusks to Foraminifera to corals changes. But whether at 300 or 3,350 feet, what are recorded are the products of carbonate organisms, separated by a short distance and an immense interval of time.

625–810 feet:	Buff, weakly cemented limestone made up chiefly of well preserved corals . . .[31]

1,975–1,978 feet:	Hard, white to cream-colored limestone . . .[32]
3,053–3,055 feet:	Chalky, white limestone . . .[33]

From 4,553 to 4,610 feet, no samples were brought up, but Ladd recorded in his log that the drilling rate remained unchanged, denoting that the material was likewise the same: limestone.

And then, at 4,610 feet, everything changed. The drill that had been streaking through time and space for several days suddenly slowed to a crawl.

With cold precision, Ladd recorded his last entry in the log for hole F-1: "4,610–4,630 (total depth) No samples; drilled with difficulty; presumably basalt."[34]

Presumably basalt.

The significance of those words was not lost on Ladd. Elsewhere in his scientific paper he writes that F-1 was "the first [hole] to penetrate the sedimentary section to the basement rock."[35]

A short time later, Ladd's team moved their drilling operation to the opposite side of the atoll, onto a slightly larger island. There, at hole E-1, they drilled down to 4,208 feet before striking a harder layer. This time they were able to bring up a sample. It was black with thin veins of sparkling calcite: olivine basalt. Volcanic rock.[36]

Writing of his discovery a few years later, Ladd concluded that Enewetak was a limestone cap built by corals (and other organic constructors), sitting on top of a basaltic volcano, itself rising two miles from the ocean floor.

"It thus confirmed," wrote Ladd, "one of the most important features of Darwin's subsidence theory. . . ."[37]

After more than a hundred years, Darwin's basic theory on the formation of coral atolls had been proved correct. Ladd wrote in understatement, as befits a good scientist. But that does not mean he was blind to irony. In his first report on the drilling operation at Enewetak, he recapped the history of atoll drilling, mentioning that Alexander Agassiz, the recipient of Darwin's "doubly rich millionaire" letter, attempted his own drilling operation in Fiji in 1897. The Swiss scientist had reached the eighty-five-foot level when he received word that the Royal Society had far surpassed Darwin's hypothesized six-hundred-foot limit for coral construction. That takes care of that, Agassiz presumably thought. He packed up his equipment and sailed home.

"It is unfortunate that Agassiz did not continue his drill hole in Fiji," Ladd points out, "as he might have reached a volcanic foundation within the limits suggested by Darwin."[38]

Ironically, it turns out that the basaltic layer is found at a much shallower depth in Fiji. Unfortunate for Agassiz, but fortunate for Ladd, who could tell his grandchildren that he was the one to prove Darwin's theory of coral reef formation.

If F-1 gave us a better understanding of the natural world, then its own sad fate tells us something equally important about human nature.

The same men who had sent the big rig to Enewetak swapped one time machine for another. This new one was very different from the drill, however. Rather than burrowing through time, the new machine obliterated it.

In September of 1952, dozens of men flew into Elugelab to build the new time machine. They were accompanied by many more men, who carried guns and stood guard around the device. Unlike the old drill rig, this time machine looked the part: it was a cylinder, with foot-thick steel walls lined with lead. Completed, it stood twenty feet high and over six feet wide, and weighed eighty-two tons. Inside the huge canister was a triple-walled stainless-steel flask containing several hundred liters of deuterium, also known as heavy hydrogen, frozen into a liquid state at approximately minus 423°F.[39]

It was the world's first hydrogen bomb, and it had been assembled in the South Pacific for its initial test. Unlike F-1, this time machine was given a true name, and a friendly one at that: it was called Mike, and it was the reason F-1 was drilled in the first place. The drilling team had been deployed to ensure that Enewetak's geology was stable enough for the H-bomb test. Proving Darwin right was mere serendipity.

At a fraction of a second before 7:15 A.M. on the first of November, 1952, Mike was detonated. It exploded with a force nearly one thousand times greater than that of the atom bomb dropped on Hiroshima. The complex chain reaction begun inside of Mike resulted in a fireball that burned hotter than the sun.

It was dawn. Coral polyps, fat with zoox, would have been slowly withdrawing into their calcium dens after a nighttime spent capturing zooplankton. Parrotfish would have been emerging from their mucus cocoons to begin their daily hunt for algae. Crabs, shrimp, and other crustaceans would have been crawling from their lairs. It was that magical moment of transition, with creatures of the sunlight replacing the many nocturnal feeders.

In nanoseconds they were all vaporized, transformed into a swirling, purplish cloud of unimaginably hot gases. This fiery mist shot twenty-seven miles up into the pale blue Pacific dawn, where it drifted with the cold winds blowing along the edge of the Earth's atmosphere.

Thousands of miles away, in a basement laboratory at the University of California in Berkeley, Edward Teller, the self-described "father of the H-bomb," saw the seismic evidence of the tremendous blast as sound waves that had crossed the Pacific Ocean in twenty minutes. He turned to a friend and commented, "That's very nice."[40]

The holocaust left behind a crater over a mile wide and two hundred feet deep.

On the map of Enewetak Atoll, where the outline of the tiny island of Elugelab once existed, there is now only an unbroken expanse of blue indicating water.

4

The Rise of Corals

The environment is the theater and evolution is the play.

—G. Evelyn Hutchinson

It started to rain in the spring of 1993 and never let up.

There were many hours in which rain didn't fall, of course, but even then our eyes followed the heavy clouds as they scudded low overhead, and we knew that it was only a matter of time. The rain washed over sodden fields and rushed into creeks that fed into tributaries that cascaded into churning rivers colored chocolate brown and thick with foam. The rivers had long ago passed flood stage. During those weeks it seemed that the distant throb of thunder was never silent for long. And every time that awful rumble sounded, we felt it in our stomachs and thought, "Here we go again." Ministers retold the story of Noah in churches across the heartland, and even many atheists began to pray for an end to the rain.

It was called "the flood of the century," although in the end it turned out to be the most devastating flood in U.S. history. All along the Mississippi River, towns disappeared beneath floodwaters. The local papers recounted the many tragic stories of people who had lost everything to the flood. Almost daily they ran a picture of someone paddling a canoe or riding a johnboat down what used to be a major street. Satellite images of Iowa looked more like a birdseye view of an immense lake dotted with islands.

"Iowa, Colored Blue," was the title the *New York Times Magazine* chose for their story on the flood. For a while it looked as if the Hawkeye State were going to join Superior, Michigan, Huron, Erie, and Ontario as the sixth Great Lake.

From late June on, the water built up rapidly behind the dam located a few miles above Iowa City. Every day the newspaper told us how many inches the Iowa River had risen behind the dam. But what really concerned us was that every day the distance separating the water from the top of the dam decreased by the same amount. The Coralville Dam is not a particularly big one, but it didn't need to be. The Iowa River is a modest waterway, with its headwaters in north-central Iowa and its terminus in the Mississippi River a short distance below Muscatine, Iowa.

When the Army Corps of Engineers built the dam in 1958, they added an emergency spillway, a wide and sloping concrete apron that in 1993 still had never been used—except by skateboarders from Iowa City. Below the spillway was a two-lane road. Below that was a camping area and picnic grounds, a pleasant wooded area filled with clumps of soft-maple and ash. Beneath these trees were thickets of mayapple, spring beauty, trout lily, jack-in-the-pulpit, and nettles where birds congregated and deer browsed—an altogether unremarkable but beautiful assortment of eastern Iowa flora and fauna.

On July 5 a cloudburst north of Iowa City finally pushed the water level above the emergency spillway. A wall of water surged over the top of the spillway and came crashing down on the land below. For twenty-eight days the water continued to pound away at this land, scouring the road and everything below it. Trashcans and the picnic tables disappeared in a swirl of dirty water and probably ended up on a snag or a sandbar somewhere south of St. Louis. Plants were flattened by the floodwaters and then washed away. As the deluge continued, saplings and then mature trees were uprooted and carried off like twigs in a brook. Still, the water continued to pour over the spillway, smashing down on land denuded of all vegetation and gnawing away at glacial-aged sediments.

When the rains finally slowed and the river level receded enough to remain behind the dam, park officials were able to reach the area below the spillway. Astounded by what they found, they called in geologists.

The flood had washed away fifteen feet of soil and clay, revealing a perfectly preserved fossilized coral reef from the middle Devonian period. It was as if a pearl had emerged from beneath layers of filth. The period, approximately 375 million years ago, is often referred to as the Age of Fishes, but that seems a dubious claim for a period in which corals reached their apex. Corals were so successful that they covered more of the planet during the Devonian than at any other time before or since.[1] They had fully developed the art of reef building, much as it's practiced today.

Still, much was different then. Time itself was different, or at least the experience of it was. Because the Earth rotated on its axis more rapidly then, a Devonian year—the time it took the young planet to complete a full revolution around the sun—was around 400 days.[2]

The very face of the planet during the Devonian, the spread of land and sea, would have been unrecognizable to us today. There was only one giant landmass, the supercontinent called Pangaea. And there was just one giant ocean, Panthalassa.[3] What was to become Iowa hovered close to the equator, drifting slowly northward beneath a shallow but extensive body of warm tropical waters.[4]

I have a piece of fossilized coral from this period. It's not much to look at: a cone-shaped rock scarcely three and a half inches long. It's light gray, nearly white at the widest end, which is two inches across. Then it darkens to a dingy black as it curves and tapers to a point. I picked it up absentmindedly a few years ago, on some weekend ramble over rock-strewn land outside of Iowa City, not far from the spillway, dropping it into my coat pocket with little thought. Today it sits on my desk, an echo of a world so far in the past, and so different from today, that it is hard to imagine that it hails from the same planet. It is the mineralized remains of a horn coral, one of a variety of solitary corals that thrived in the Devonian waters. Three hundred seventy-five million years ago (give or take a few million years) the dark, pointed end was anchored in the sea floor. At the top, the widest part, was the coral's mouth, and from there radiated delicate tentacles, capturing their microscopic prey in much the same way corals do today. Horn corals were solitary organisms, although the order they belonged to, Rugosa, included colonial members. Although they probably lacked symbiotic algae, the Rugosa represented a significant step forward for corals, since they were the first members of the group to form skeletons.[5] By the time my specimen was munching on plankton in tropical Iowa (a descriptor of the state that my wife repeats to herself longingly during our long winter months), the Rugosa had already been around for quite some time, possibly something on the order of 100 million years.[6] In fact, they may have already passed their peak as a group and were heading downhill, albeit slowly. Barely surviving a worldwide extinction event at the end of the Devonian, they were reduced to a few individuals and then slowly made a comeback. But it was for naught. The Rugosa met their end during the Permian crash[7] (around 245 million years ago), an apocalyptic unparalleled event that destroyed up to 96 percent of marine animal species.[8]

In addition, another order of colonial corals, the Tabulata, were widespread in the area (and throughout the world) during the Devonian, and fossilized mounds of it were also found in the reef exposed by the flood. The Tabulata are significant because they may have been the first corals to develop a symbiotic relationship with zooxanthellae.[9] They, too, barely survived the Devonian crash, only to be wiped out by the Permian extinction.

❄

But I may be giving a false impression of the historical role of corals, for the words *reef* and *coral* have not always been interchangeable.

Assume for a minute that you are a detective, or a scientist—who is surely a detective of sorts. Assume that your evidence includes only corals living today and the Devonian fossils found below the spillway in Iowa. What would you conclude has been the sole reef builder in the history of the planet? With just the evidence at hand, you'd probably answer, "Corals." But you would be wrong. In fact, corals are just one of several organisms that, over the long history of the planet, have built reefs. True, today they have the market on reef construction nearly cornered. But at one time or another many other organisms have filled this niche. Algae and bacteria, bivalves, foraminiferans (unicellular marine organisms with hard shells), bryozoans (tiny colonial creatures that resemble moss), and sponges have all had their moment in the sea over the long stretch of geologic time.[10]

The first reef builders, using the term *reef* loosely, were cyanobacteria acting in concert with microalgae to trap sediments and erect dense mats of these materials on the sea floor.[11] The resulting mounds, called stromatolites, were only a few yards high. They first appeared in the fossil record a couple of billion years ago and continue to grow in some parts of the globe.[12]

But those "reefs" bore little resemblance to the complex communities we know today. A step closer was taken by archaeocyanthids, possibly primitive cup-shaped sponges, which trapped calcium-rich mud, about 560 million years ago.[13] At least those "reefs" were home to other simple organisms, such as trilobites, the now-extinct but once-ubiquitous arthropods that have the distinction of being the first animals to evolve a high-definition eye.[14]

The first reefs worthy of the name, however, didn't evolve for another 50 million years or so, and were based on sponges called stromatoporoids. A young German doctoral student named Gert Wörheide is the champion of a probable descendent of these former stars, coralline sponges of the genus *Astrosclera.* I talked with him in 1996, after he had presented a paper on these "living fossils."

"The stromatoporoids were the dominant reef builders at the beginning of the Phanerozoic [about 570 million years ago]," he told me, with enthusiasm. He explained how these early sponges—the very first multicellular animals—formed skeletal spines called spicules, made of calcium. "They used to be the king of the reef," he said. The tall young scientist reminded me of a father boasting about his child making the honor roll or being selected as captain of the soccer team.

"But the world changes," Wörheide said with a philosophical shrug. As the environment changed, other organisms reigned. By the Jurassic (208 million years ago) stromatoporoids were "only accessory elements" on the reef, said Wörheide, pushed out by corals. Even later, the *Astrosclera* sponges were pushed off the reef surface entirely, and into cryptic habitats—mostly in caves where zooxanthellate corals couldn't survive. The precocious child had become the ne'er-do-well adult, eking out an existence on the margins. Today the main ecological role of sponges in general on the reef is as a destroyer, churlishly (to ascribe human emotions to the process) undermining the reef built by its vanquisher by boring into the limestone foundation.[15]

Corals are certainly the dominant reef builders today, and it's true that they have existed in one form or another for over 500 million years. But if you were to trace their rise to the top, you would end up not with a straight line, but rather with a series of sharp ups and downs. They have even come precariously close to total extinction several times. Scleractinian corals, or "hard corals," the modern reef builders, don't appear in the fossil records until around 245 million years ago.

Just as *coral reef* may be misleadingly used to refer to reefs of all types, the word *coral* itself is the source of some confusion. *Coral* is the name generally given to members of the order Scleractinia, the stony reef builders. There are many other corals, however, and perhaps it's best to place these corals in the proper context of their cousins and various relations.

Imagine a large umbrella labeled Cnidaria. Huddled beneath are corals, jellyfish, sea anemones, and a few other organisms. (You may hear members of this same group referred to as coelenterates, but don't let that confuse you; the terms *cnidarian* and *coelenterate* are interchangeable.) These organisms share certain physical characteristics. Most important of these are bodies composed of two cell layers, the presence of stinging cells, and the fact that they are radially symmetrical—which simply means that if you cut the organism in half, through the center at any point, you will have two mirror-image segments.

Beneath the umbrella labeled Cnidaria, imagine another, smaller umbrella. This one is labeled Anthozoa. Forget about jellyfish and the other miscellaneous animals; they have been left behind.

Which leaves us with corals and sea anemones.

Now imagine two smaller umbrellas, side by side beneath the one called Anthozoa. But here's a surprise. According to taxonomists, these umbrellas are not labeled *sea anemones* and *corals*, as you might expect. Instead, one umbrella is for Octocorallia (and includes many "soft" corals), and the other is labeled Hexacorallia and includes both sea anemones and the hard corals, the scleractinian reef builders.

As the Greek roots of their names implies, this division is based on the fact that octocorals have eight tentacles and hexacorals make do with six.

Several varieties of octocorals, in a group called gorgonians, do not secrete true limestone skeletons, but these colonial corals do produce small bits of calcium carbonate found throughout their soft bodies. Many of them also have a flexible central skeleton made from a protein-based material called gorgonin. While these delicate creatures are relatively rare throughout the world, they are widespread in the Caribbean, where they are an important part of the reef community. Gorgonians take on a dizzying variety of shapes and colors. There are lacy brown soft corals, thick stands of yellow, bushy candelabrums, swaying plumes of sea feathers of all colors, and, most impressive of all, sea fans, which live in veritable fields like huge flattened flowers, purple or yellow, always facing at right angles to the current so that they can filter out as much plankton as possible.

But it is the hexacorals, and in particular, the scleractinians, that we probably think of when we think of reefs. Even under this smaller umbrella labeled Scleractinia, the diversity is astonishing. Approximately seven hundred separate species of scleractinians have been identified, and there are no doubt more awaiting discovery.

Even within a single genus, the "umbrella" just below Scleractinia, different species can assume a vast variety of shapes. They can form thickets of branchlike colonies. They can build boulders several yards in diameter, or form fat knobs that grow only a few inches. They can erect massive tabletops that appear lacy up close or grow in delightful spiraling whorls. Or they can simply encrust a hard surface with a formless spreading mass.

One genus is particularly adept at disguising itself: *Acropora*. It is the coral of a thousand faces. All right, perhaps not a thousand. But it can assume any of the shapes described above, and there are far more species within the genus *Acropora* than in any other. So far, the tally is set at 150 "confirmed" species.[16] I put the word in quotes because taxonomists are constantly debating whether a new specimen constitutes a truly unique species or whether it is merely a new form (or morphotype) of an existing one. Until a consensus is reached, these morphotypes remain in taxonomical limbo where they are labeled "nominal species." If we include the nominal species in our count of *Acropora*, the total is far higher: up to 368[17]—or more than half of all identified coral species.

Acropora is the superstar of the coral world, and by the standards of geological time it was an overnight sensation. Like the understudy who rises to stardom when the lead falls ill, the emergence of acroporids resulted largely from the misfortune of its competition. *Acropora* originated sometime in the Eocene Epoch (between 58 and 37 million years ago), and for many millions of years it remained just another voice in the reef chorus. Then, about two million years ago, just before the first true humans walked onto the stage, catastrophe struck: two-thirds of all coral species went extinct.[18]

And there in the wings stood *Acropora*. The genus spread so rapidly into the recently freed ecological niches that a coral scientist has termed this period "the Acropora Revolution."[19]

Acroporids (members of the genus *Acropora*) have a number of competitive advantages, what biologists call a suite of adaptations, which allowed them to spread quickly and dominate coral reefs, particularly in the Indo-Pacific region. Two of those characteristics are particularly important.

I "discovered" one of them as a student at a marine laboratory in the Florida Keys. It was during a summer intensive course on coral reef ecology, and our instructor, a soft-spoken but intense coral physiologist named Erich Mueller, was teaching us how to break living samples of *Acropora palmata* (commonly called elkhorn coral) into smaller pieces that would grow into new microcolonies—if we didn't mangle them too badly in the process. It was a difficult process for neophytes, made harder by the fact that it had to be accomplished underwater. We each took a turn, climbing onto a plastic stool and reaching down into the giant aquarium where a large branch of *A. palmata* with many smaller lobes sat, defenseless. The trick was to nip the bottom of a lobe in just the right place using a pair of bone cutters, so that the new microcolony would snap off—and then catch it with your free hand before it drifted to the bottom.

That was the theory, anyway. In practice, each of us gnawed away ineffectually at the coral, nipping off tiny bits. The more macho students would fasten on to a chunk that was far too large and squeeze as hard as they could—again accomplishing nothing. There is probably no better exercise to bring home the fact that the *Acropora* skeleton is made of stone.

When at last a lobe would snap off and we'd catch it, each of us was amazed: even accounting for the buoyancy of the water, holding the microcolony was like holding nothing at all in your hand.

The *Acropora* skeleton is a rock with the weight of a feather. It is made of the same form of calcium carbonate as other corals, but it is honeycombed with hollow chambers. That is a tremendous evolutionary advantage. It allows for a very light yet astonishingly strong structure, one that can be built very quickly, beating out its slower-growing competitors. *Acropora* growth rates are truly remarkable, with some species able to develop more than ten inches of new material in a single year.[20]

But there is another reason for the success of "the Acropora Revolution." Acroporids discovered the benefits of "division of labor" long before the Industrial Revolution made the practice widespread among humans. Specialized "axial" polyps form the growing tips of the colony, with "radial" polyps handling reproduction.[21]

Taken together, these characteristics allow for a high degree of integration among individual coral polyps, and the formation of many intricate structures, suited to a variety of environments.[22] There is an *Acropora* to fit nearly every one of the structural categories listed above. Thanks in large part to these two characteristics, acroporids, in all their many guises, cover 80 percent or more of the surface area of many reefs today.[23]

❄

As I've said, the natural history of corals has been a long, bumpy ride. Periodic mass extinctions have been the rule over geologic time. Corals flourish and evolve, adapt and spread, filling new niches, branching off new species, and then, just when they seem to be hitting their stride—BANG—some cataclysm nearly wipes them out. They disappear from the fossil record entirely for millions of years, surviving as isolated groups of just enough individual colonies to struggle back. Many species and even genera don't survive and are known to us only as fossils (such as my horn coral).

But in the short term, relatively speaking, corals have not had an easy time of it, either. Even many divers who spend a good deal of time and money to see corals up close, however, seem unaware of this fact. They'll vacation on that wondrous chain of structures called the Great Barrier Reef, and return with amazing tales of coral colonies that are millions of years old.

Peter Sale is so tired of hearing those stories he could scream. He blames Charles Darwin for this widespread misunderstanding.

"Darwin started it all with his description about reefs being built slowly," says Sale, leaning across his plate of calamari. "The *words* are true, but the *impression* isn't."

We are dining under a slowly revolving fan on the patio at Jimmy's, a pricey restaurant where they serve beer in frosted mugs and platters of fresh seafood hauled from the eastern Pacific, which washes up against a seawall only yards away. It is midsummer in the tropics, and the air is sultry. Not far away, just out in Panama Bay, ships await the green light signaling their turn to enter the huge Miraflores Locks and begin the fifty-mile passage northeast through the Panama Canal to the Atlantic Ocean.

We are both attending the Eighth International Coral Reef Symposium, the quadrennial gathering of coral reef scientists. This time it's held in Panama City, a fitting site, since it is home to the planet's second-largest biogenic structure (after the Great Barrier Reef), the Panama Canal. Sale is here to present a paper titled "Biogeographic Variation in Reef Fish Trophic Structure: Assembly Rules Differ Between the Virgin Islands and the Great Barrier Reef." I'm here scratching my head while trying to decipher titles such as that, and to interview people like Sale.

We were introduced on the Internet, on the "coral-health list-serve," an electronic conversation among six hundred coral scientists—and at least one freelance writer/eavesdropper. I knew the reputation journalists enjoyed among scientists ("'No comment' is the only quote they can be counted on to get right—and sometimes they'll screw that up"), but a few days before the symposium I had inflated my courage enough to post a message to the entire group asking if any scientists were willing to get together in Panama to talk about their work. Sale was one of the generous half-dozen who said yes. Actually, to be accurate, what he wrote back was, "I'd enjoy a conversation with a dreaded media person."[24]

We meet in one of the crowded hallways at the behemoth ATLAPA conference center and walk across the even larger parking lot to Jimmy's. Because I miss my family, I'm looking forward to Sale's company at dinner as much as I am to any information he might provide about coral reefs. Sale looks like a younger version of Jack Kemp, with the same thick salt-and-pepper hair and sharply chiseled features. To me, he is just another conference participant. It isn't until much later that I learn he is revered in the field, the editor of *The Ecology of Fishes on Coral Reefs*, a volume considered to be the bible on the subject. It is a good thing I didn't know any of this at the time, or awe might have spoiled a fascinating dinner conversation.

Sale complains about Darwin when, between mouthfuls of grilled octopus, I ask him about popular misconceptions concerning coral reefs.

"Oh," he says wearily, "all that stuff about 'the oldest known ecosystem.' It's part of the mystique." It's accurate as far as it goes, he adds, but it's misleading. Sale has developed an antidote to this simplistic view. He uses it on his first-year students back at University of Windsor in Ontario. In a darkened room, he shows them a slide taken on the Great Barrier Reef, highlighting the dazzling diversity of corals and marine life there. There are the predictable oohs and aahs.

Then he projects an image of the Great Pyramid sitting in the Egyptian desert.

Silence.

"The second structure," he informs his students, " is older than the first."

This is always followed by an even longer period of silence, punctuated by the squeaking sound of students shifting uncomfortably in their chairs.

But it is a fact: the Great Pyramid was built some 4,500 years ago.[25] The corals living today on the Great Barrier Reef did not reach the surface until approximately four thousand years ago.[26]

We have left behind the abstract and somewhat surreal realm of geologic time. *These* changes have taken place on a time line that is most definitely of human scale.

You can't be too hard on Darwin, however. He had no idea that these coral reefs were of such recent origin. The theory of subsiding landmasses was controversial enough in his day. The notion that climatic change could cause significant changes in sea level was not even widely discussed back then, and wouldn't be for several decades.[27]

But the main point isn't who understood the effects of climatic change first; it is that the many changes in the level of the world's oceans, caused by periods of glaciation and melting, have had a profound effect on the formation of modern coral reefs.[28]

It is one of the many ironies concerning corals that in modern times the fate of these tropical creatures has been largely dependent on what has occurred at the polar extremities.

This doesn't imply that sea-level change caused by glaciation is a recent phenomenon, however. Glaciers have been accumulating at the poles and melting away for many millions of years, affecting the amount of water available to the world's oceans. There is plenty of fossil evidence to indicate this. It is true, however, that in the middle of the Miocene Epoch, approximately 16 million years ago, the world entered a "glacial mode"[29] during which these "up-down" periods have become far more frequent. Even more important is the fact that geological factors of subsidence and continental drift take place slowly. When you look at a relatively short period of time, sea-level changes jump out as the factor controlling reef growth. This is especially true in areas that are comparatively inactive geologically—areas such as the coast of Australia and the Caribbean.

The basic principle is straightforward: as the Earth cools, more water accumulates at the polar ice caps in the form of snow and ice, and as a result, the sea level drops. As the Earth warms, the process is reversed and water comes flooding back into the sea. The technical term for the episodes when the newly replenished oceans cover what was dry land is "transgression" (more evidence that science is done by terrestrial beings—as if the oceans were unjustly invading *our* territory).

For corals, a change in either direction can spell new-found success or, more often, death. If the sea level increases slowly, many corals are able to take advantage of the new water and expand upward. Today's mean sea-level rise of about two inches per decade[30] fits within this range and is within the growing abilities of many corals, particularly members of *Acropora*. But if the flooding occurs too quickly, coral reefs can drown. That's what happened to a great many reefs between ten and twelve thousand years ago, during the most recent great glacial melting event. During that period the ocean rose more than sixty-five feet in less than two thousand years (or nearly four inches per decade, twice today's rate). Some coral reefs managed to survive by a neat little trick called backstepping, shifting into reverse and growing onto rubble or any higher stratum that might be available.[31]

At the Fifth International Coral Reef Symposium, held in Tahiti more than a decade ago, two scientists introduced a new way of looking at reefs, based on sea-level changes. They classified reefs by their "reaction" to rising sea levels, identifying three types of reefs: keep-up, catch-up, and give-up. Keep-up reefs are found in shallow waters and are composed of fast-growing branching corals that match sea-level increases. Catch-up reefs form in deeper water with more massive (and slower-growing) coral boulders. As sea levels rise, branching corals take over and the reef survives, its coral composition changed. Give-up reefs simply stop accreting, usually because they can't keep up with the rising sea level.[32]

A dramatic *drop* in the ocean level can be far deadlier for corals. The giant Wisconsin glacier that covered large portions of the Northern Hemisphere between eighteen and twenty thousand years ago sucked up so much water that the sea level dropped 443 feet below its present point.[33] Corals that had been established off Australia were left high and dry. This was the era of land bridges throughout the world. Australia, Papua New Guinea, and Tasmania were all parts of a single landmass.[34] Alaska and Siberia were connected. Mainland Asia was connected to what is today the western part of the Indonesian archipelago.

It wasn't until nine thousand years ago, when the seas approximated their present-day level, that modern corals reefs began slowly to reestablish themselves,[35] on the foundations laid by their ancestors. As Peter Sale points out, it took several thousand years more to develop the complex reef communities now found in places such as the Great Barrier Reef.

So although reefs were among the very earliest biogenic structures on the planet, coral reefs appeared a bit later in the game. And although coral reefs have existed for hundreds of millions of years, *living* coral reefs are nearly all relatively young, their life span measured in the thousands of years, not in the millions.

But does this take away any of their mystique? On the contrary. Coral reefs are at once ancient and recent. If they are a braid, then they are a braid through time, weaving together events of the dim geologic past with those of the present and (with some luck) of the future. As a group, corals have taken everything nature can throw at them and rebounded time after time. Falling oceans, rising oceans. Sliding continents and closing seaways. Drastic climate change. Perhaps even a collision with an asteroid. It hasn't mattered. For all their delicate appearance, corals as a group are tenacious, resilient, and inventive.

Compare the tiny, gelatinous coral polyps with the fearsome dinosaurs, with their massive bodies, armorlike hides, immense claws and teeth. *Tyrannosaurus rex* is gone. But the humble coral remains.

5

The Heart of Lightness

The World Calling on Dawn

Come, O my dawn!
And dawn on me!
Come, O my morn!
And gaze on me!

—Traditional chant, Papua New Guinea

We are stretched out on the rear deck of a motorized *prahu*, a traditional Indonesian wooden boat, so that Sergio has to shout to be heard over the roar of the engine: "They say when Manado Tua wears a hat in the morning, it will rain in the afternoon!"

He rises onto one well-tanned arm and points to this weather-predicting mountain and shrugs. Sure enough, a "hat" of clouds obscures its peak. But it is still early, and at least for now the sky is a brilliant blue and the sun is shining. It glints off the polished surface of Manado Bay with an intensity that is stunning. Sergio Cotta is a young marine biologist who, along with three other Italian expatriates, runs a dive shop in the city of Manado, on the northeast tip of the island of Sulawesi, Indonesia. Their decision to locate here has everything to do with the amount of solar energy pouring down on us as we head to our dive sites.

The tropics are by definition sun-drenched. Look at a map of the Earth: there are two dotted lines girdling the planet at approximately twenty-three and a half degrees on either side of the equator. The line to the north is called the Tropic of Cancer; the one to the south is the Tropic of Capricorn. The area between these lines is called the tropics—where the sun is directly overhead at midsummer.

Even in the tropics, the closer you get to the equator, the more sunlight there is (measured in kilocalories per square meter). Manado is just a single degree of latitude above the equator. At noon there are no shadows here. There is only the blinding light.

Solar energy of this intensity can be threatening to those of us high up on the food chain. "Stay out of the sun, especially at midday!" shriek guidebooks, and with good reason. The intensity of the equatorial sun can redden exposed flesh in minutes and blister skin in an hour. The sunlight can awaken sleeping cancer cells. The old saw "only mad dogs and Englishmen go out in the noonday sun" comes from this region.

But to the true sun-worshipers of this planet, the photosynthesizers, this is the Holy Land. It is their land of milk and honey, where the raw material for food—sunlight—pours down in ceaseless abundance. Coral reefs—among the most productive ecosystems on the planet—are uniquely adapted to take advantage of this intense light.

Before you put your map away, turn to the area between Asia and Australia. Locate the Philippines and place a dot there, on the northern tip of Luzon. Then go south and west until you find the Indonesian island of Sumatra. Put another dot there. Now head due east until you find Papua New Guinea and place a dot there. Finally, connect the dots with a straightedge. What do you have? A giant triangle straddling the equator.

Enclosed by that triangle is arguably the most important region on Earth. It is the planetary center of biodiversity.

Here, life on land and sea runs riot, fueled by vast amounts of high-octane sunlight and pushed on by a number of other flukes of geology, ocean currents, and other fortuitous factors. When I asked one scientist what makes this region so special, he just stared at me, bug-eyed, groping for an answer that would do justice to the magic triangle. He finally blurted out, "This is where it's at!"

By "it," he meant life, of course.

There is nothing modest about life within this triangle. On the contrary, life here is a shameless braggart, constantly reminding you of its own inventiveness, shouting from every direction, "Look at me! Look at me!" On the island of Bali, I sat on my hotel porch one night for an hour, bottle of Bintang beer in hand, watching as a multitude of bizarre insects paraded by, attracted to the light, each one seemingly more unusual than the last.

But in Manado it is the coral reefs I have come to see, and the undersea world found in this triangle is at least as fantastic as what exists topside. If you were to trace a line around the area in which the greatest number of genera of corals are found, you'd end up with another triangle that nearly perfectly duplicates the one you've already drawn. The major difference is that this new line dips down at one corner to include the Great Barrier Reef. Within this line there are seventy known genera of corals.[1] (Pity the poor Caribbean, which hosts only twenty.)

It is a forty-five-minute trip across Manado Bay out to our dive spot at Tanjung Pisok. The channel here is not very wide, only a few kilometers across, but it is extremely deep—close to a mile down in places. As we skim across the water, pleasantly groggy with the light, I think about Alfred Russel Wallace, the great naturalist and explorer who, independent of Darwin, arrived at the theory of evolution through natural selection. For Wallace, the large and oddly shaped island of Sulawesi (it has been likened to an enormous, tipsy letter *K*) was crucially important. He spent more than a year here, including several months traveling in the area around Manado, where he kept a small house and which he described as "one of the prettiest [towns] in the East."[2]

Outside of Manado, it is still an area of great beauty, with several volcanic peaks rising above graceful coconut plantations. Manado itself is now noisy and dirty and my hotel, a former Dutch mansion, went to seed long ago. The only interesting things about it are the lizards that streak up and down the walls during the night and squeak exactly like mice.

At any rate, it wasn't the beauty of the area that most impressed Wallace, but the fact that nowhere else on his extensive travels throughout the Indonesian archipelago did he find so many species peculiar to a single island.[3] The importance of Wallace's sojourn in Manado is suggested by the fact that he sent the first manuscript on his theory of natural selection to Darwin immediately after leaving the town.[4] And it was the delivery of that manuscript that prompted Darwin to come forward with his own theory, one that he had been sitting on for years, afraid to publish for fear of the inevitable attacks from religionists. But with Wallace having reached the same conclusions—and prepared to publish them—Darwin decided he had better make his own theory public or lose out on any credit for developing it.[5]

Wallace doesn't mention the corals of Manado in his writings, but marine creatures weren't his specialty anyway, and it's quite likely he never saw them. If he had, Wallace probably would have marveled at the reefs there as he did later at those around Ambon Island, a few hundred miles to the southeast. (See chapter 1 for his description of those corals.) One of Wallace's contemporaries did, however, explore the reefs of this area: Sidney Hickson, another British naturalist/explorer. He tried to write about the reefs here as a naturalist does, with long lists of flora and fauna. Finally he threw up his hands.

"A coral reef cannot be properly described," Hickson concluded in 1889. "It must be seen to be thoroughly appreciated."[6]

Those are precisely my thoughts after we arrive at Tanjung Pisok, don our scuba gear, and dive into the water. What possible combination of words can adequately convey this new world just beneath the waves?

Part of the overwhelming sense of awe elicited by reefs comes from the suddenness of the transformation. One moment the oceanic world is flat and nearly featureless, except for the brilliant dance of light on water. And then, like Alice stepping through the looking glass, you stride off the boat and break through that silvery surface. Instantly, you are in a different world. The only line more definitive and mysterious than the one separating the world of air from that of water is the boundary between life and death.

Beneath the sea, you are surrounded by myriad creatures, some of them attached to rocks and waving in the current, multi-hued fishes swimming about in huge glistening schools, crustaceans scurrying in and out of countless holes. So many organisms. So many colors. So many unknown creatures flitting here and there. Especially to the newcomer, the reef appears chaotic in the extreme. It's like being dropped into the center of a huge city on an alien planet. But life on a coral reef is not truly chaotic; it is complex, and the difference between chaos and complexity is called ecology. Like taxonomists, ecologists are in the line-drawing business. The difference is that while taxonomists draw lines to separate organisms from each other, ecologists draw lines to connect them.

The analogy between a reef and a city is anything but original. It may have come from Darwin, but no one seems to know for sure. One relatively early reference comes from an ecologist named Eugene Odum. It was Odum who in 1955, along with his brother Howard, wrote one of the most important papers on coral ecology, based on research the pair conducted at Enewetak Atoll (again, in conjunction with the U.S. nuclear weapons program).[7] The Odums partially solved the dilemma of how corals survive on what seems to be a starvation diet. According to one biologist, their work "produced some of the most challenging and creative hypotheses concerning the coral reef since Darwin . . ."[8]

Eugene Odum went on to author one of the early important textbooks on the science of ecology, the study of the interrelationships between organisms and their environment. In explaining how ecosystems work, Odum drew upon his earlier experiences at Enewetak, using the example of a coral reef, which he called "one of the most beautiful and well-adapted ecosystems to be found in the world."[9]

A city is like any other ecosystem. It may appear chaotic to the newcomer, but if you study it carefully, drive around and get to know it, the chaos settles into patterns—complex ones to be sure, some of which even defy our current powers of understanding, but patterns nonetheless.

Drive through any large city in the United States. What do you see? Approaching from the countryside, you will likely pass through suburbs. The people who live there are often engaged in similar activities; they are "professionals" who labor with their minds instead of their hands. Most are white. In swankier sections you'll find huge houses and similarly outsized lawns. If you poke your head into the attached multi-car garages, you're likely to find late-model foreign cars, minivans, and well-appointed four-wheel-drive vehicles. Keep driving. As you continue, incomes decline, and so does the size of the houses and lawns. Cars are older. People here are more likely to do manual labor. Continue on and you may pass through an industrial area. It may be filled with rusting cranes and tumbledown brick buildings, or the old buildings may have been replaced with newer ones made of glass and aluminum. Eventually you arrive in the inner city. There are no lawns here. (Although, if the section is affluent, it may have central green spaces.) If the area hasn't been gentrified, you'll find older buildings, many small shops, and poor people of different races and from various countries living side by side.

Reefs, too, have their neighborhoods, each with its own special physical characteristics and populations. The principle is called zonation, and just as no two cities are exactly alike, each reef is unique. Zonation is markedly different on a Caribbean fringing reef from that found on a Pacific atoll. But, again like cities, most modern coral reefs share enough characteristics that scientists can make some generalizations about zones.

Take, for example, the fringing reef at Tanjung Pisok. Like most others, it is divided into two major sections: the fore-reef and the back-reef. Separating those two zones is the reef crest. (Some researchers lump the crest in with the back-reef; others claim it is more properly part of the fore-reef. Scientists have been known to come to blows over just this sort of thing, but laypeople have the luxury of not taking sides and just noting that the reef crest does exist.)

This is zonation at its most simplistic—it's like dividing human geography into rural and urban areas and leaving it at that. But urban areas comprise many sub-zones, and so do reefs. In cities, this stratification is largely controlled by money. On reefs, other forms of power define zonation: waves and intensity of light (the latter is dependent largely on depth).

We approached Tanjung Pisok in the same fashion as our driver in the example above approaches the city—from the outlying areas inward. Here the equivalent of the countryside is the deep oceanic water off Sulawesi, the domain of pilot whales, sharks, and porpoises. At some point approaching Tanjung Pisok, in waters so deep that they have yet to be explored by humans, the deep fore-reef begins, where the bottom begins to slope upward toward the surface.

Of course we don't begin our dive here. You need special equipment to withstand the tremendous pressure at this depth. Besides, as terrestrials we start from the top and go downward. So our tour of Tanjung Pisok begins at the zone of transition between the back-reef and the fore-reef, just seaward of the reef crest. The reef crest itself—the highest point of any reef—is typically ignored by divers. Pummeled as it is by waves that few but the hardiest corals can withstand, exposed to air at low tide, there's generally not much to look at here.

On many reef crests there is an algal ridge, where a particular kind of red algae covers nearly every inch of rock and rubble. This is especially true where reefs are regularly battered by powerful, storm-produced waves, such as Caribbean reefs in the path of hurricanes.[10] Coralline algae are the unsung heros of the coral reef. Their vital role in reef construction has been overshadowed by the more exotic lifestyle of their cousins, the zooxanthellae living inside corals. But coralline algae play a role nearly as important in the construction of reefs. Just like scleractinian corals, these simple algae have the ability to remove calcium from seawater. The coralline algae construct their cell walls from calcium carbonate—the same material that forms coral skeletons.[11] As they encrust coral rubble, coralline algae fuse it into a solid mass of stone. They are the glue that holds the nonliving reef together—and, by providing a stable substrate, allow the living reef to continue its upward growth.

Just seaward of the reef crest, where we begin our dive, the fore-reef begins its slow, sloping descent into deeper waters. Here, where solar energy is optimal, many terraces are covered with lush stands of corals. They are mostly acroporids, but some fifty-eight genera of corals have been counted in these waters,[12] and a few of these others are mixed in with the more dominant genus. Fish, most of them small, dart around the corals. Sergio taps my arm and points to a petite, strikingly colored fish that stays close to a colony of branching corals. The front half of this fish is a bright, almost electric purple. Halfway down its slender body, the color changes abruptly to an equally brilliant yellow. Later, Sergio informs me that this small fish, with the big name of *Pseudochromis paccagnellae*, is very rare outside of these waters.

On the windward side of many reefs, where wave energy is at its greatest, there is a distinctive feature known as a spur-and-groove formation.[13] Here, long spurs, or ridges, of rock covered with corals or algae alternate in a more or less regular fashion with deeper grooves through which water and sediments flow. Often the pattern isn't obvious to the diver, who may be too close to the subject to notice the regularity of this formation. But from the air, spur-and-groove architecture is one of the first things you notice about windward reefs.

There are no abundant spurs and grooves at Tanjung Pisok. Instead we find steplike terraces of corals. As we descend, the quality of light changes. Here in the heart of lightness, the actual intensity of illumination remains relatively unchanged for a surprisingly long time. Colors, however, undergo a much faster metamorphosis. One by one colors are extinguished, beginning with those of the longest wavelength. Descending is like going gradually color-blind. Red is the first hue to go, absorbed by the seawater in just a few meters. Gone are all the brilliant shades of scarlet, crimson, and ruby. Red's neighbor on the spectrum, orange, holds on a little longer, and then it, too, drowns.

At about sixty-five feet, coral species diversity reaches its maximum level. Acroporids are no longer able to crowd out the rest. More than a hundred feet below the surface, we level off and head northward, parallel to the reef crest. Gone, now, are the large stands of branching acroporids. At this depth, the very beginning of the lower reef slope, the diminished light results in fewer hard corals (dependent on internal sun-gathering zoox), and so soft corals and other animals become more common. One of the most striking residents of this subzone (at Tanjung Pisok) are the many colonies of "spiral wire coral." Each colony resembles a giant yellow corkscrew, several feet long, that appears to spring from the reef slope. Looking at it, you can almost hear the *b-o-i-n-n-n-n-g-g-g*. If Dr. Seuss had been commissioned to design a reef animal, the "spiral wire coral" would have probably been the result. Despite its outer color, the spiral wire coral belongs to a group known as "black corals"—named for the color of their skeleton. Instead of producing calcium carbonate, black corals build a skeleton from a tough protein called "horn," which is hard and very nearly black.

Even at this depth, and despite the relative scarcity of hard corals—and the fact that colors are muted—the abundance of life is still staggering. Mostly it is the cast of characters that has changed. Besides the many species of black corals, there are soft corals and sponges everywhere. The temptation is to continue swimming, in a greedy effort to see more and more. But this is a foolish notion—a habit based on terrestrial, temperate experience. The secret of diving in the heart of biodiversity is to stay in one place, to hover and focus on what's directly in front of you. You'd run through your entire tank of air before seeing all the different organisms in that one spot. Besides the black corals, there is an abundance of soft corals, sea fans, sponges, and crinoids.

Crinoids are a particularly beautiful class of creatures, the most ancient relatives of sea stars and sea urchins, all members of the large phylum Echinodermata. Rachel Carson called the echinoderms "the most truly marine" of all invertebrates, because not one of its approximately five thousand species has ever made the transition from the sea to land; none even lives in fresh water.[14] Over 500 million years ago, even before corals had appeared, crinoids lived in these deep waters, their flowerlike arms radiating from a central head and covered with delicate projections called pinnules. Attached to the reef substrate by a single stalk, these crinoids, largely unchanged over the passing eras, are commonly called sea lilies.

Sponges are also plentiful at this depth, including many as large as a person. These simple animals are so ancient that they make even the crinoid seem like a recent invention. Sponges are, in fact, the most primitive of all multicelled animals, appearing in the fossil record millions of years before more-complex animals formed. These ancient members of the animal kingdom also have the dubious distinction of being an evolutionary dead end. They probably evolved from single-celled animals (protozoans) between 800 million and 1 billion years ago and then—nothing.[15] Almost miraculously, no major forms of organisms have evolved from this phylum over the ensuing time. For some reason this experiment in multicellularity did not lead to bigger and better things. Instead, life went back to the drawing board and started the experiment over again with a different lineage, one that includes corals.

But we should give credit where it's due, and sponges have been very successful in other ways. They've shown amazing staying power and species richness: approximately five thousand sponge species have been identified, and the true number may be far higher.[16]

I'm still focused on this one small section of the enormous reef when Sergio taps my arm and indicates that I should check my pressure gauge. I am running low on air, so he gives the sign to surface.

Stripped of our gear and changed into shorts and T-shirts, we eat lunch on the nearby island of Bunaken. We sit on a shaded balcony overlooking the ocean, where we are served plates of meat, rice, and cooked papaya leaves. Dog meat is a specialty in Manado, so I sniff suspiciously at the plate, my mind guiltily conjuring up images of my beagle, Buddy, at home. Finally, Sergio smiles knowingly and says, "Chicken." (He later tells me the sad story of how he made the mistake of bringing his dog with him from Italy when he moved here. When the animal disappeared, he asked around, but his neighbors just shrugged and avoided his eyes.)

After lunch, Sergio and I stroll along the beach, the brilliant equatorial sunlight pouring down almost as tangible as rain. Squinting out toward the water, I can just discern the reef crest, where the surf breaks, leaving a white tail of foam. Just inside this ragged line begins the other large zone: the back-reef. By far the most extensive subzone of the back-reef is the reef flat. It begins on the landward side of the reef crest and stretches away from the sea anywhere from a few hundred feet to a mile or more. Close to the reef crest, sheltered from the pounding oceanic waves by the reef crest, a great diversity of corals, fish, and other animals thrive. (It was this subzone, at Heron Island on the Great Barrier Reef, which I described in greater detail in chapter 1.) But as you continue back toward land, corals and fishes both become less abundant in this shallow water. Strips of sand, the result of ground-up corals and coralline algae, cover the bottom with ever greater frequency. The limitations on upward growth in the area combined with increased sediment in the water produces an intriguing and beautiful structure called the microatoll.[17] The inner reef flat at Tanjung Pisok is dominated by these microatolls. Each one is built by a single colony of the coral *Porites*, forming a living circle of coral between six and fifteen feet in diameter, with a pool of water in the center. The tops of the colonies are smothered by sand and other sediment and die. Gradually the dead coral erodes away, leaving the central pool, which is home to many other creatures. The inner portions of some of these microatolls have been radiocarbon dated at between five thousand and six thousand years old.[18]

And here, according to the conventional wisdom of centuries past, is where the coral reef ends.

After our stroll along the beach we head back to the *prahu* and then chug a short distance to our last dive site of the day, a place called Likuan, off the southernmost edge of Bunaken island. The fore-reef here is very different from the one we explored earlier at Tanjung Pisok. Rather than gradually descending in a series of coral-covered terraces, the reef at Likuan plummets suddenly into the immense depths separating the island from the mainland. It is known as one of the best "wall dives" in the world, with canyons and smaller cracks in the reef face, sheltering innumerable creatures: corals, sponges, sea snakes, turtles, and the beautiful and improbable tunicates, sometimes called sea squirts. Tunicates can take a variety of forms and colors, but they're generally shaped like miniature vases, only a few inches high, with a smaller opening on the side. Like sponges, tunicates are filter feeders, pumps that draw in water through their larger hole and drive it back out through the smaller one, after having removed food particles and oxygen. I call them improbable because, for all their simplicity, they are our distant relatives, members of the phylum Chordata, a name that refers to a nerve cord. Our spinal cord is the most developed form of this structure, surrounded by protective backbone. The tunicate's cord is at the opposite end of the spectrum. It doesn't even exist in adults. But a larval tunicate, which resembles a tadpole, has a notochord (a hollow tube) running down its back that contains a functioning nerve cord. It doesn't function for long, however. The larval state lasts only hours, and by the time the creature has metamorphosed into an adult like the ones on the wall at Likuan, this primitive nervous system has melted away.

The wall at Likuan also teems with fishes. A 1981 study of the area by the World Wildlife Fund described the site as "spectacular," and diving there, it's not hard to see why. Clouds of silver fusiliers swirl around schools of the oval yellow-and-white pyramid butterflyfish. Dozens of other species of fishes dart in and out of corals.

As we descend, those fishes are replaced by other species. Even at a hundred feet, creatures of all kinds remain abundant. Hovering for a moment, I gaze down into an immense darkness that is punctuated occasionally by quick flashes of bluish light. An unlikely association comes to mind: of camping at night in the desert outside Tucson, Arizona, bewitched by the silent blue streaks of lightning from far-off thunderstorms rolling east over the mountains in Mexico. But here on the other side of the planet and a hundred feet underwater instead of in the desert, the flashes are the result of the last remnants of sunlight glinting off the scales of fishes that live in those dark and cold depths. Looking up is even more compelling. It's like staring into a shimmering blue sky that, as if in a dream, is filled by great flocks of birds with iridescent wings moving in unison and in slow motion: more fishes.

Too soon, our air supply runs low and we are forced to rise into the quicksilver heavens.

I'm still intoxicated by the profusion and the beauty of life below, and so when at last we break the surface, leaving behind the world of water and entering into the realm of air and land, I barely notice that the aphorism Sergio told me earlier about Manado Tua "wearing a hat in the morning" has proved true. Deep into the heart of lightness, we ascend into rain.

6

The Outer Strands

No man is an island entire of it self.

—John Donne, Meditation XVII

John Donne was a great poet and philosopher, but he would have made a lousy marine ecologist.

No *island* is an island—at least not in the sense that Donne meant it: "entire of it self."

Coral reefs had long been considered "islands" entire of themselves, and there was compelling reason for this point of view: their high primary production (the conversion of sunlight into food for other organisms), combined with their efficient nutrient recycling within the reef itself, have led many scientists to consider reefs as entities apart from their surroundings. In fact, when the ecologist Eugene Odum first compared reefs to cities, it was chiefly in this context, commenting on their *dissimilarity*.

"We note again," wrote Odum, "that the city . . . but not . . . the coral reef, is a heterotrophic [referring to those organisms which cannot manufacture organic compounds from inorganic sources] ecosystem dependent on large inflows of energy from outside sources."[1]

Well, yes and no.

Certainly it is true that coral reefs are oases of life in the nutrient deserts of surrounding oceanic waters. But a growing number of scientists are realizing that the concept of coral reefs existing in isolation is greatly, and dangerously, oversimplified. Even the extent of "tight nutrient recycling" on the reef, which was unquestioned for many decades, is now subject to new investigation.[2]

Coral reefs themselves, for all their complexity, are a single strand in a larger braid—a braid within a braid. Another strand in this larger braid is an overlooked ecosystem that is fascinating and beautiful in its own right: the sea-grass meadow.

Not every coral reef has a meadow of sea grass nearby, but most do, and those meadows or beds play an important, though still not well understood, role in the life of the reef. Like most recreational divers and snorkelers, I had taken sea grass for granted—or worse, as an obstacle one must swim over before reaching the "good stuff" on the reef.

But Tony Larkum teaches me otherwise.

We share a dugout canoe, piloted by a Kuna Indian, poking around the reefs of the San Blas Islands, off Panama's Caribbean coast. We had already shared a breathtaking, if somewhat unnerving, predawn flight in a small plane from Panama City on the Atlantic side, over the isthmus's rain forest, setting down on an island landing strip that was barely long enough, the plane coming to a halt only feet from the water's edge. Once we're safely on board the canoe, I strike up a conversation with Larkum, a native of London who has spent the last three decades at the University of Sydney, Australia. His soft-spoken, unassuming, yet dignified manner, is engaging. In our short boat ride, I learn that while Larkum mostly studies algae now, he used to be quite involved in the study of sea grass. I casually mention that I'd like to discuss his work later, and he agrees.

Our guide drops anchor by the Aguadargana reef complex and we jump into the water. I quickly lose myself perusing a single massive coral head, larger than I am and nearly covered with Christmas Tree worms, delicate filter-feeders that construct calcareous tubes tunneled deep inside the coral, into which they withdraw their twin red-and-white spiraling tentacles when threatened. Every hole on the coral head is filled with sea urchins and a bevy of other creatures, and my entire time in the water is taken up circling this one coral head.

Later our canoe returns us to our base camp on Nalunega Island, and I settle into a beach chair and enjoy one of those perfect

Caribbean days on which the wind blows just enough to keep the bugs off and the sun feels so warm and sensual on your flesh that you almost believe you are photosynthesizing. Behind me I can hear the *molas*, the Kunas' exquisitely embroidered cloths, tied to a line and available for purchase, flapping in the breeze. (Traditional patterns depicting sea creatures share space with such modern motifs as Beavis and Butthead.) I am drowsing in this blissful state when a shadow falls across my eyes. I open them to find an excited Tony Larkum standing by my elbow, his white hair dripping cold sea water onto my knees, his hands full of sea grass.

"Look," he commands, thrusting some blades beneath my nose. "*Thalassia*."

For the next half hour I get an intensive lecture on the natural history, importance, and aesthetic qualities of sea grass, particularly of this variety with the euphonious name *Thalassia testudinum*, more commonly known as turtle grass. By the end of his lecture, I, too, am smitten by this deceptively simple plant.

Here's the abbreviated version: One fine day (speaking in evolutionary terms), primitive plants marched out of the sea. Progress. They developed a vascular system. More progress. They obeyed God's commandment: Be fruitful and multiply—the key word being *fruitful*. Fruits contain seeds, the culmination of flowering. These flowering plants are called angiosperms, literally, "seed in a vessel." The method proved so successful that flowering plants soon dominated the land.

Just as angiosperms were beginning to take over on dry land, an ancestor of *Thalassia* apparently grew homesick and marched back into the sea, probably beginning with tentative forays into the saltwater marshes of Cretaceous Europe, long before the first large mammals appeared. It was among the first life-forms to make the great trek homeward. Many millions of years later, marine mammals such as porpoises and whales followed a similar path—from ocean to land and back to the sea.

One doesn't simply go from land to sea in a single jump, of course. Thomas Wolfe had it half right: You *can* go home again—only not quickly, and it takes a hell of a lot of work. So much work that sea grasses are the only land plant to have made the transition. *Thalassia* and its relatives developed an entire suite of adaptations to survive in the old neighborhood.

Sitting on the beach in Panama, dripping sea grasses in hand, Larkum points out one of the most elegant adaptations. Salt water is corrosive to most leaves, but *Thalassia* developed mature leaves that are saltwater-tolerant.

"The problem," he says as he sits down beside me, "is that the young leaves aren't able to survive in salt water." The plant had to find a way to protect young, tender leaves from salt water, while growing in that medium. Larkum holds a collection of oval-tipped blades only a few inches long, which disappear into a single sheath, about an inch and a half long. With a surgeon's skilled and gentle touch, he delicately separates the blades from the sheath, revealing a pale green immature leaf inside. The leaf is surrounded by another transparent sheath.

"Push on it," whispers Larkum, indicating the sheath.

I prod it with my index finger. It's cushiony, like my infant son's fluid-filled teething toys.

"Water," says Larkum, a delighted smile playing across his face. "Fresh water."

The plant bathes the infant leaf in water it has desalinated and bottled up in this protective sheath. When the leaf has matured enough to stand up to the rigors of salt water, it grows out of the protective sheath, leaving behind another young shoot.

Like most other sea grasses, *Thalassia* spreads by two methods: sexually, by flowering and producing seedlings; and asexually, by an underground stem, or rhizome, which grows horizontally. The rhizome sends thick roots down and new shoots up at regular intervals. The roots form a mat extending down several feet, adding tremendous stability to what might otherwise be shifting sands. The sea-grass blades act as speed bumps for currents and waves, slowing them down and trapping sediments.[3] The leaves grow remarkably fast, up to three inches in a week,[4] and can provide a dense cover of more than three thousand blades in a square yard.[5] (The leaves are generally flat, but in the afternoon, when photosynthesis is at its peak, they produce more oxygen they can readily release, and the blades swell up as thick as fingers.[6])

The protection against erosion provided by sea-grass meadows is just one of their many important contributions to coastal ecosystems. "If you took away the sea grasses," says Larkum, "you wouldn't have this"—and with a sweep of his arm he indicates the small island where we're sitting. "Or those," he adds, pointing to the mangrove trees growing in the brackish waters straddling land and sea. "You almost always find coral, mangroves, and sea grasses together," he muses.

Larkum has, in his offhand manner, introduced another important strand in the braid we're considering: mangroves.

These three ecosystems—corals, sea grasses, and mangroves—have been intimately associated for over a hundred million years, back to the time when the Tethys Sea separated the two supercontinents of Laurasia and Gondwana.[7] Through time and space, as continents broke apart and wandered over the face of the Earth, these three ecosystems have become ever more tightly bound together, a relationship that scientists have only recently begun to plumb.

Coral reefs are revered, and deservedly so. But like "charismatic mammals," such as the panda or the elephant, they are easy to love. They are the "charismatic ecosystem" of the tropics. Sea grasses are underappreciated at best, and at worst considered a nuisance, despite their ecological importance.

But mangroves are another story altogether. Westerners have always treated mangroves with contempt and even outright hatred.

Consider this assessment by an early New Zealand novelist:

> Oh! those mangroves. I never saw one that looked as if it possessed a decent conscience. Growing always in shallow stagnant water, filthy black mud, or rank grass, gnarled twisted, stunted and half bare of foliage, they seem like crowds of withered, trodden-down old criminals, condemned to the punishment of everlasting life. I can't help it if this seems fanciful. Anyone who has seen a mangrove swamp will know what I mean.[8]

That may seem harsh, but it is typical. Most Westerners, coming originally from the temperate zone to the tropics in search of riches, found mangrove forests—thick and impenetrable, rife with alligators, crocodiles, and diseases—a formidable obstacle to conquest. Even Charles Darwin, who so often saw beauty where others did not, had little good to say about this benighted tree:

> The channel . . . was bordered on each side by mangroves, which sprang like a miniature forest out of the greasy mud-banks. The bright green colour of these bushes reminded me of the rank grass in a churchyard: both are nourished by putrid exhalations; the one speaks of death past, and the other too often of death to come.[9]

Frankly, if you have ever tried to walk through a mangrove forest, you probably agree with these men. It is a thoroughly unpleasant and nearly impossible chore. (The world's record for the hundred-meter dash through mangroves is twenty-two minutes, thirty seconds.[10]) Mangroves often grow with their roots sunk deep into stinking muck, and this can be a wonderful breeding ground for mosquitoes carrying dengue fever and malaria. Admittedly, mangroves are not the aesthetic equivalent of coral reefs. Still, mangrove forests play a vital role in the life of coral reefs, and support a striking variety of life in themselves. Fishes, shrimps, and especially crabs are abundant, with up to eighty individual crabs scurrying around a single square yard of mangrove forest.[11]

And mangroves do have a quiet beauty all their own, as some writers have observed. Alison Lurie wrote appreciatively of the "low gray-green mangrove islands float[ing] on the horizon like vegetable whales."[12] Rachel Carson also treasured this rich ecosystem, describing the mangrove forests surrounding the Florida Keys as "silent, mysterious, always changing."[13]

My wife and I once paddled a kayak through the area Carson visited. Following a large southern stingray over a lush sea-grass meadow, we then turned into a narrow channel that meandered through an island of red mangroves. Instantly we were in a different world. Extensive "prop roots" rose from the water, joining the main trunk a foot or two higher up, giving the impression that the trees were standing on stilts. A canopy of long, waxy leaves dappled the intense subtropical sunlight as we drifted along, dreamlike, sheltered from the wind and waves. Hanging from many trees were pendulous, cigar-shaped seedlings, which actually begin to sprout while still on the tree, dropping into the water to drift many miles (hundreds, sometimes) before finding a new spot to colonize. Spiderwebs were everywhere, connecting leaves from different trees. There were smallish webs built by crab spiders and the more magnificent structures erected by the huge yellow-and-black golden orb weavers. Occasionally we'd catch sight of a bird, usually a brown pelican or a cormorant drying its wings, hidden among the foliage. More often we'd only hear them and many other birds—great white herons, egrets, frigate birds—all twittering, clucking, and cooing from their leafy redoubts.

Mangroves do not form a distinct taxonomic group; that is, they aren't genetically related to each other in the way that oaks or maples are. They form a group based solely on the fact that they fill a particular niche—in the same way that lawyers or writers may be considered "family," although they are not related by blood. Historically, between 60 and 70 percent of tropical and subtropical shorelines have been lined with mangrove forests[14]—for a rough total of 122,000 square miles.[15]

Like sea grasses, mangroves have adjusted to a specialized environment with a suite of adaptations. Most important is their ability to live in salt or brackish water. Mangroves have accomplished this task in a number of ingenious ways. In the mangrove forests of the Florida Keys, for example, there are three types of mangroves, each with a different system for dealing with salt. The red mangroves my wife and I paddled through are beautiful trees with dark red wood beneath gray bark. They can exist either in small, shrublike clumps or, under the right conditions, grow up to eighty feet high.[16] Of the three mangrove species growing in the Keys, the red mangroves live the farthest from land and are therefore exposed to the highest concentrations of salt. A living example of "preventive medicine," the tree uses a form of reverse osmosis to block salt from entering its roots. Still, some salt does invade its vascular system, and so the cell walls of the red mangrove leaves are capable of extruding the mineral.

The black mangrove cannot prevent salt from entering its system, so it specializes in efficiently eliminating the substance through leaves covered with special pores through which salt is dispersed. (The Indians living in the Keys before the Spanish arrived used this to their advantage, scraping salt crystals off the leaves to add to their food.) The white mangrove, which lives the farthest from salt water, can make do with twin salt-eliminating pores found at the base of each leaf.

The putrid mud mentioned by Darwin smells like rotten eggs because it is full of hydrogen sulfide, an indicator that there is very little oxygen in the soil.[17] To obtain oxygen, mangroves may "breathe" through small raised pores (called lenticels) on their prop roots, or through fingerlike projections (pneumatophores) that grow out of the water surrounding the tree. Even relatively small trees, ten feet or less in height, may produce more than ten thousand pneumatophores.[18]

This muck is also an important clue to the mangrove's relationship with coral reefs. Mangroves filter sediments and pollution coming from land, providing a buffer zone between the terrestrial and marine environments.[19] Mangroves also absorb nutrients coming from land, passing some along to sea-grass beds, which use them and in turn pass a fraction on to the reefs. By gradually reducing the amount of nutrients in the water, mangroves and sea-grass beds keep reef waters low in nitrogen and phosphorus, preventing algae blooms that could smother corals.[20]

Boil all this information down, and you are left with an essential fact: Mangroves and sea-grass beds are the two lines of defense protecting sensitive coral reefs from potentially devastating land-based influences.

Coral reefs, sea-grass meadows, and mangroves: each one is an important, and linked, coastal ecosystem. But physical zonation of the kind described thus far, while important, is only one aspect of a system—and probably one of the less informative ones. It's like describing a city by listing the buildings you find in each area and ignoring the inhabitants of those structures. I've touched briefly on some of the flora and fauna, but even that misses the main point. What makes reefs (and sea-grass meadows and mangrove forests) so interesting isn't just their biodiversity, but how individuals and groups there *interact.*

When I was a boy, there was a popular television program called *Naked City*, set in New York City. All I remember of the show was that it dealt with the sometimes violent and sometimes poignant interactions among the inhabitants of that great metropolis. At the end of each episode, a disembodied voice would intone, "There are eight million stories in the Naked City. This has been one of them."

There are far more stories on the coral reef, and, just as on *Naked City*, they cannot all be told—certainly not in a nontechnical book such as this one. But a few of these stories should be recounted, in part because it is the dynamics that make coral reefs the fascinating and unique places that they are. The relationships between the teeming organisms on the coral reef are among the most complex of any on the planet. Coral reefs are the Russian novels of the sea world, full of passion and avarice, convoluted and interweaving story lines, and colorful characters by the dozens. And of course these stories are full of sex and violence, for—with apologies to prudes and the tenderhearted—such is the way of the world.

7

A Song of Love and Death

The man best fitted to observe animals . . . would be a hungry and libidinous man, for he and the animals would have the same preoccupations.

—John Steinbeck, *The Log from the Sea of Cortez*, 1941

It was the spring of 1979, and twenty-six-year-old Jamie Oliver's thoughts turned toward sex. Unlike many of his classmates at Australia's prestigious James Cook University, however, the object of Oliver's attention wasn't human: he was fixated on the reproductive habits of the staghorn coral, genus *Acropora*.

His adviser told him to forget it. Some of the top marine biologists in the world had been studying this question of coral reproduction for years. No one was quite sure how *Acropora* reproduced, and Oliver was informed that he—a mere graduate student, after all—was unlikely to solve the puzzle. But Oliver couldn't forget it. On the contrary, what had started as an interest evolved into an obsession.

The tall, lanky Canadian had arrived in tropical Townsville, Australia, in 1974 after studying biology at McGill University in Montreal. He had planned on spending just two years "down under," but when he was accepted into JCU's honors program in marine biology, he decided to stay. His timing was fortuitous. In the mid-1970s, a handful of researchers around the world were independently beginning to reevaluate long-held notions about coral reproduction, and Oliver was young and brash enough to believe that he, too, could become a trailblazer in this new frontier of research.

A few maverick researchers doesn't add up to a revolution, however. The dogma surrounding coral reproduction was as thick and impenetrable as the mangrove forests on tropical shores. Researchers had discovered early on that corals reproduced both asexually, by budding (polyps cloning themselves), or sexually. The mechanism of budding was fairly straightforward and most scientists had believed that sexual reproduction was nearly as uncomplicated. For a century, conventional wisdom held that corals sexually reproduced *internally*, with sperm fertilizing eggs within the polyps.[1] According to this theory, the resulting zygote grows inside the polyp until it reaches maturity as a larva. Eventually the full-grown larva, or planula, just a couple of millimeters long, is released through the polyp's mouth. This process of internal fertilization followed by a live birth is known as viviparity or brooding. Once free, the planulae swim and drift their way to a new home, where they settle onto the reef bottom and start a colony.[2]

There was nothing wrong with the theories of viviparity and "planulation," as far as they went. Some corals do follow this pattern. The problem had more to do with the way science is done (i.e., by humans, who are fallible). Planulation was discovered early, around 1790, and by the turn of this century, theory had hardened into dogma.

"There is little doubt that all corals are viviparous," declared one eminent scientist in 1905.[3]

From that edict, things went from bad to worse. Research revolved around particulars of brooding and planulation. How, exactly, did they occur? Where in the polyp? When? Those were the only questions to be answered. And when researchers were faced with a species that refused to produce planulae, they abandoned the coral and moved on to one that was more "cooperative."

Yet, as far back as 1859,[4] a few scientists observed something very different occurring in some species. Those researchers were swimming against the scientific current, however, and were easily ignored.

But over time, the sheer number of "uncooperative" coral species began to attract attention. Some scientists started asking why, if all corals were viviparous, more of them didn't show it. Put another way, where the hell were all the planulae? There was a convenient reply, however. The accepted explanation was that since corals could reproduce sexually *and* asexually, if planulae weren't found, it merely suggested that most corals reproduced asexually.[5] That answer sufficed, even into the late 1970s. It was simple, logical, and wrong.

In 1979, Jamie Oliver broke off several pieces of *Acropora* coral and saw small, round, pink structures. He carried more coral pieces home and examined the globules under a microscope. There was no doubt about it: they were oocytes, developing egg cells. The acroporid was preparing to spawn.

Excited about his discovery, Oliver eventually found others in his program who were thinking along similar lines. A group of six of them—four men and two women—started meeting to discuss the issue. They were each studying a different coral species, or different aspects of the same species, but they had a lot in common: they were fascinated with coral reproduction, and they were bright, young, and ambitious.

"We weren't stellar," says Oliver today, "but we were good enough to recognize that *something* was going on."[6]

Soon they found themselves in the middle of a fascinating scientific adventure. They bounced theories off each other, rejecting some outright and holding on to others for further inquiry. They pored over existing literature, looking for flaws in methodology and logic, and for clues to other possible avenues of reproduction. They found a paper containing a tantalizing suggestion: "Even though viviparity is considered to be the rule among corals, it is possible that some species which have not been found to release planulae may release both eggs and sperm which fuse in the water."[7]

They were also influenced by Alina Szmant's work on a nonreef coral growing in the temperate waters of Narragansett Bay, Rhode Island. Back in the summer of 1970, Szmant was a member of the first team of female "aquanauts," a group of five scientists who spent two weeks in an underwater laboratory conducting experiments on a reef in the Virgin Islands. Despite all the silly and sexist publicity surrounding the women's group (some press reports referred to them as "aquababes"[8]), they were all serious scientists. Out of the public eye, Szmant, like the others in the group, had gone on to do important scientific research.

The article that caught the eye of the young Australians was one of Szmant's most important papers, the first report detailing the formation and development of gametes in a coral. What interested the Australian group most was Szmant's discovery that *Astrangia danae* was a "broadcast spawner," that is, it released eggs and sperm into the water column. She also speculated that reef corals could be spawners as well.[9]

Other scientists identified some corals as spawners, but these were still viewed as anomalies, the exceptions that proved the rule of brooding. By the end of the decade, corals known to brood still outnumbered spawners by a ratio of four to one.[10]

In the spring of 1980, some members of the JCU group were diving on a nearby reef when they noticed maturing oocytes in several coral colonies. When they looked again a few days later, the eggs were gone. They had to have spawned, the students reasoned. But when and how?

Finally, in the spring of 1981, the group placed several corals in a large aquarium and waited for the spawning. They were like nervous fathers in a hospital waiting room. The bundles formed, full of the raw materials of sexual reproduction, and wrapped in a pink sheath as if in a baby blanket. The corals appeared ready to spawn. But nothing happened. The bundles remained firmly in place. The hours passed and the group became more and more discouraged. Finally the exhausted students turned out the lights and went to bed. When they returned the next morning, the bundles had been released.

The lights had been the key. The corals, like some of their relatives, were waiting for darkness to spawn. The group set up several aquaria filled with different coral species. In addition, they were monitoring corals on three separate reefs. As the spring sun warmed the ocean waters, changes began to take place in the corals they were observing. Oocytes and sperm-egg bundles grew and gradually took on color (usually pink). The group knew that spawning and planulation were sometimes linked to lunar cycles, and so they waited for the full moon. On October 13 the moon rose, full and glowing, over the ocean east of Townsville.

Five nights later, on October 18, the spawning that the students were awaiting finally began. Three coral species spawned. They knew that one, *Acropora formosa*, had spawned because they found its gametes floating in the aquarium in the morning. Spawning was also inferred in a similar fashion in the case of a second coral from a different family, *Goniastrea aspera*.

Things were different for *Goniastrea favulus*. A fine-mesh bag covering an individual coral on the reef was found the next morning to contain egg-sperm bundles. But there was another specimen of *favulus* in an aquarium, and that night the students, for the first time, actually observed the spawning process. Polyps shot forth a stream of sperm, followed by a sticky mass of eggs. In a single night the students had increased the number of species known to spawn by nearly 40 percent. On the next night, four additional species were observed spawning in aquaria, and one for which spawning had been inferred was now seen ejecting its pink egg-sperm bundles. In two nights, they had nearly doubled the number of species known to spawn. Five nights after the next full moon, on November 16, some of the students witnessed corals spawning in the field, on a reef just a few miles offshore Townsville. Six more species were added to the list of spawners, for a grand total of fourteen.

But the most important discovery was still a year off.

At sundown, on November 6, 1982, members of the JCU team donned their scuba gear and dived into the water. As dusk descended, first one coral colony and then another began to spawn, polyps releasing their pink egg-and-sperm bundles into the water column, where they floated to the surface and broke apart, the sperm fertilizing the bobbing eggs. Everywhere the students looked, colonies of corals of different species were filling the reef waters with their gametes. So many bundles were shooting from so many polyps simultaneously that the gametes formed a huge slick on the surface. One scientist later observed that it was like being caught in an "underwater snowstorm."[11]

What they were witnessing was an annual synchronized mass spawning, an event so momentous that it's hard to imagine how it wasn't widely known much earlier (others had, in fact, noticed these spawning events, but had never published their findings). Eventually it was determined that at least half of all the billions of corals up and down the entire Great Barrier Reef, stretching for more than a thousand miles, spawn at dusk, near the fifth day after the full moon in the late spring.[12]

The JCU team's findings appeared as the cover story in *Science* magazine[13]—quite a coup for a group of graduate students. Their first and most obvious achievement was to shatter the orthodox view that all corals were viviparous.

"As a result of this study," they wrote, "more coral species are now known to spawn gametes than to brood planulae, suggesting that viviparity may be the exception, rather than the rule in coral reproduction."[14]

Researchers around the world began to investigate this phenomenon more closely. (Others, such as Alina Szmant, had already been at work looking for spawners in Caribbean reef corals.) By 1986, 167 coral species were known to spawn, while only 60 brooders were recorded.[15] The latest figures available indicate that 85 percent of coral species studied are spawners.[16]

But the discovery of synchronized mass spawning was at least as important, for it explained some important aspects of coral success, while raising new and exciting questions. Synchronizing spawning gives corals several evolutionary advantages. Because the water is filled with these gametes, predators are likely to have their fill without seriously depleting the eggs and sperm. By releasing the bundles simultaneously, eggs stand a better chance of being fertilized, if not by the sperm of their home colony, then by those from a neighboring one.

The fact that many different species synchronize their spawning gives scleractinian corals another important evolutionary edge: hybridization, the crossing of species boundaries. Bette Willis, a member of the original JCU team, has specialized in this area, exploring the hybrid zone—the place where "interacting species meet and mate."[17] The most intriguing part of this research is the possibility (some say the probability) that such hybrid zones may be "windows on the evolutionary process." By studying how different coral species cross-fertilize during episodes of mass spawning it may be possible to actually witness that engine of the evolutionary process: the birth of a new species. Botanists have long understood the role that hybridization plays in the evolution of plants; the discovery of coral mass spawnings presents zoologists with the challenge of determining whether hybridization plays as important a role in the evolution of this branch of the animal kingdom.

Mass spawning also forced scientists to look at reefs in an entirely new way. Brooded planulae generally find a place to settle and colonize in a relatively short period: between a few hours up to a couple of days, depending on the species.[18] When most corals were thought to brood, it was easy to assume that reefs self-seeded—that is, that new colonies were the offspring of other colonies on the same reef. This reinforced the notion of coral reefs as self-contained units.

But planulae produced by spawning don't have the physiological head start provided by brooding—it takes some spawned species several days before they're capable of settling.[19] And even when capable of settling, these larvae can survive for weeks before finding a new home.[20] Given this new information, the role of ocean currents takes on tremendous significance, serving as watery highways for coral larvae.

This was a radical theory when Richard Grigg, a zoologist from the University of Hawaii, first proposed it in 1981—three years before the Australian team's *Science* article. Grigg and a couple of other scientists had only recently established the modern existence of *Acropora* in Hawaii, at a place called French Frigate Shoals, in the center of the archipelago. *Acropora* had been found in Hawaii's fossil record, but had gone extinct locally millions of years ago. With the discovery of living *Acropora*, scientists had to consider how the genus was reintroduced. One possible explanation was that the coral planulae hitched a ride on the bottom of James Cook's ship in 1778. That famous British explorer was traveling from the coral-rich regions of the South Pacific when he became the first European to set foot on the Hawaiian Islands. It was an intriguing theory, but Grigg and his colleagues ruled it out after analyzing core samples from the Hawaiian reef that showed that acroporids had been growing there for more than one hundred years *before* Cook's arrival.[21]

Grigg found many other explanations equally implausible. But the idea that planulae drifted over on an ocean current also faced some large hurdles. First was the question of currents themselves. Humans have been studying ocean currents longer than we have any other marine phenomenon, and for good reason. They've been *our* watery highways for millennia, allowing trade between continents and mass migrations of whole peoples. The ancient Polynesians certainly had a sophisticated understanding of the Pacific currents. And since Cook's time, Western science has added to this knowledge with extensive charting.

Since the 1940s, oceanographers have known that the primary current affecting the Hawaiian islands is the North Pacific Equatorial Current (NPEC). The problem for Grigg's theory of coral recruitment by drifting planulae was that the NPEC flows from east to west—and there are no significant reefs east of Hawaii containing acroporids and close enough for their larvae to survive the trip. There are reefs with *Acropora* conceivably within range of brooded larvae, at a place called Johnston Island, but the problem is that they are 450 miles *southwest* of French Frigate Shoals. The NPEC would carry larvae spawned from Johnston Island away from, rather than toward, Hawaii.

Once again, being terrestrial creatures has limited our thinking. The currents understood by the Polynesians and charted methodically by the British were all surface currents, generated by the sun and wind. Understanding those currents may be adequate for those traveling by boats on the surface. There is far more to the ocean, however, than what is found on its "skin." In 1951, oceanographers discovered a tremendous countercurrent flowing beneath, and in the opposite direction to, the surface current in the southern Pacific Ocean.[22] Such a countercurrent running from west to east beneath the NPEC was first hypothesized in 1969, and its existence was confirmed in 1978. The Subtropical Countercurrent (SCC) flows from Japan all the way to the Hawaiian archipelago.[23] Grigg admitted that, given the speed of the SCC and the known lifespan of brooded planulae, his theory that modern Hawaiian

Acropora came from Johnston Island was a bit shaky. Still, it was the most plausible explanation going. But when the Australians showed that most acroporids were spawners, with long-lived planulae capable of making the voyage eastward on the SCC, Grigg's theory suddenly went from plausible to probable.

Whether or not spawned planulae have a greater viable range than their brooded counterparts remains a controversial question,[24] but this may be moot in one respect. In the 1980s, scientists discovered that adult corals catch rides on whatever happens to be drifting by (masses of seaweed, pieces of wood) and make those same transoceanic journeys to new homes, a process known as "rafting."[25]

The result of these new discoveries is that coral reefs can no longer be considered independent, self-perpetuating ecosystems. Undoubtedly there are reefs that exist in splendid isolation; a few Pacific atolls come to mind. But, from what is now known about coral reproduction and recruitment, it is quite possible that the fate of many (if not most) reefs depends upon the existence of other reefs hundreds, perhaps thousands, of miles away. Once again the metaphor of a braid seems appropriate to distant reefs—separated by vast, empty stretches of water, but woven together by oceanic currents.

❊

Once a coral reproduces (whether sexually or asexually, through brooding or spawning), the fight for life is far from over. It has just begun. Corals face many threats, including from each other. Family feuds are always the most vicious, and because corals are sessile animals—tethered to the bottom—competition over space is a primary and never-ending battle. Corals from the genus *Acropora* have partially solved this problem by growing fast enough to overshadow their rivals. This form of competition, in which no physical contact is made between the competitors, is termed "indirect." It's similar to how terrestrial plants, such as fast-growing trees, crowd out other species by shading them from life-giving sunlight.

But in the past two decades ecologists have made huge leaps in understanding another form of competition between corals:

"direct" battles between these seemingly passive creatures. Contact between the soft tissues of two coral colonies is like the bell that begins a boxing match. The contestants come out swinging. This analogy goes only so far, however, for though boxers must rely on their gloved fists, corals have a surprising variety of weapons at their disposal. In some cases it is literally an eat-or-be-eaten contest, with coral polyps extruding their digestive filaments through their mouths and tossing them onto their neighbors, where enzymes dissolve the enemy polyps.[26]

Other corals use the same stinging tentacles, or nematocysts, that they employ to capture plankton. The two colonies lob stinging barbs at each other, like artillery shells from opposing armies, until the more venomous coral compels its foe into retreat.[27]

A more advanced variation on this theme is the development of specialized tentacles called "sweepers"—several times longer than normal, and tipped with an increased number of powerful nematocysts. Some species develop sweeper tentacles only after contact has been made with a rival.[28] Once the battle has been won and the sweeper tentacles are no longer needed, they are withdrawn and, in some corals, atrophy.[29] Still other corals—particularly soft corals—engage in a primitive but highly effective form of chemical warfare, releasing toxic compounds into the water where they injure or kill neighboring competitors without touching them.[30] And some corals simply produce extra mucus that they use to coat their neighbors' tissues—sliming them into oblivion.[31]

Scientists have only begun to look at these mechanisms, and more study is sure to produce new insights into how corals compete, and cooperate, to produce a reef community. But the information gained in the last decade or two has already revolutionized how we think about reefs. The concept of larval recruitment, in particular, has made a significant impact on reef studies. An optimist would say that if you destroy a single reef, another one across the globe may eventually revitalize it with new larvae. True. But the opposite is equally true: Destroy a single reef and not only is there the chance that it will never return, but that another one across the globe could *also* vanish. There is a new concept in ecology, based on this possibility, called "extinction debt." What the term means is that a perfectly healthy reef, full of living corals, fishes, and all the other species that scurry and flash around it, might be doomed—might, in fact, already be dead in an evolutionary sense.

With its supply of larvae ended, a "thriving reef" is like a clock, wound tight once and then abandoned. For a while, all seems well. But deep inside, the spring slowly unwinds with every tick, its stored energy dissipated. At some point the clock begins to slow down, its hands hesitating a fraction of second, and then more, between each movement. And when the spring has no more energy to give, when all potential energy has been converted into kinetic energy, the clock stops. Almost certainly there are some reefs today facing a similar fate. They, too, appear fine, their hidden "spring" of existing life driving them on. But, without new larvae to replenish them, those reefs are already running down. Someday, scientists or the people who live on them will detect problems. Perhaps fishermen will notice that their catch is diminishing. Or an ecologist will detect that algae has moved in where corals previously thrived. Many theories will be promulgated and debated. Conservationists will put forth solutions, plans will be drawn up. But in the end it will be for naught, for no matter what is done, eventually those reefs, like the clock left unwound, will stutter to a halt.

8

Fish Stories

THIRD FISHERMAN: Master, I marvel how the fishes
live in the sea.
FIRST FISHERMAN: Why, as men do a-land; the great
ones eat up the little ones.

—William Shakespeare, *Pericles, Prince of Tyre* (act 2, scene 1)

Not far from the pink-hued medieval buildings of Verona, Italy, where Romeo courted Juliet, is a small mountain village called Bolca. It isn't noted for its food or its wines. There are no Roman aqueducts or ancient stone bridges to attract tourists, and no epic battles were fought here. From its perch atop the Monte Lessinia plateau in the foothills of the Italian Alps, Bolca does have a pleasant enough view of the pasturelands of the Alpone River valley, but that in itself isn't enough to attract visitors. Few tourists have ever heard of this quiet town. Yet you need merely whisper the name Bolca to a fish paleontologist and his face will light up immediately, in the same way that a gourmet will smile when you say Tuscany, or an art historian at the mention of Florence. That's because the hills surrounding the village are home to an amazing collection of fossils known simply as "the Bolca fishes."

It is hard to overstate the importance of the Bolca fishes. They represent an invaluable baseline from which nearly all evolutionary research on modern reef fishes begins. The fishes are exquisitely preserved in buff-colored limestone, like flowers carefully pressed and dried between the pages of a book. Other deposits may be rich in skeletal fragments, but the Bolca fishes were fossilized whole, with skeletons and delicate fins perfectly intact. Miraculously, many of the fossils even retain their scales, still showing patterns of coloration. An eighteenth-century Irish naturalist traveled to Italy to see the fossils for himself and was so impressed with their "uncommonly perfect" condition that he wrote his colleagues that the Bolca fishes might almost be considered "fish-mummies."[1] The sheer diversity of fishes found around Bolca is itself staggering, with fossils from 247 species represented.[2]

For ecologists, the Bolca fishes are most important because they provide a fossil record of an entire reef-fish community. Some catastrophe (probably a massive algae bloom) killed all the fishes in this area at the same time, providing a crisp snapshot of the reef during the Middle Eocene age, approximately 50 million years ago, when coral reefs dominated large sections of the Tethys Sea.

The striking thing is that there is little difference between this fossil "snapshot" and a real one of coral reef fishes today. All the orders of fishes found at Bolca, save one, are still living, and 85 percent of the families still inhabit reefs.[3] Plants and animals have evolved dramatically since the Eocene. The first grasses were just coming into existence then, and the earliest hominids didn't appear until millions of years later. But coral reef fishes as a group have changed remarkably little. Such stasis hadn't been the rule, even among fishes, in the period leading up to the Eocene, when fishes evolved rapidly, with whole groups dying out and others taking their places. But in the Eocene reef, fishes found morphologies, or body shapes, that worked well, and by and large they've kept them ever since.

With a total of more than four thousand species of fishes occurring on coral reefs around the world,[4] it is a bit foolhardy to make generalizations about body types. Still, there are some commonalities. If you're swimming on a reef observing fishes, you'll probably notice three things right off: most of them are small, generally under a foot long, and often far smaller than that.[5] You won't see many "wide" fishes;[6] most reef fishes are so skinny that they appear nearly two-dimensional. And, of course, you'll immediately notice their gaudy colors.[7]

Like nearly everything about coral reefs, the evolutionary advantages provided by these features are still debated. Some tentative answers appear reasonable enough, however. Small fishes are able to exploit the craggy and labyrinthine reef environment. "Laterally compressed"—the scientific term for "skinny"—fishes have greater control over body movements, allowing for precision turns at close quarters,[8] and may be harder for predators to swallow; their great height-to-width ratio could cause them to get stuck in a predator's throat.[9] (This last point was graphically illustrated recently when a Louisiana man attempted to swallow a friend's live tropical fish. The fish lodged in the man's throat and he choked to death.[10])

The intense coloration of reef fishes is probably the most striking and controversial feature. It's certain that coloration serves a number of purposes, most revolving around those two fundamental preoccupations, food and sex. Beyond that rather general statement, you dive into the realm of scientific debate, waters nearly as contentious, complex, and fascinating as the reef itself. First comes the business of eating and avoiding being eaten (I'm now back to discussing reef fishes, but this statement could apply equally, if metaphorically, to scientists and controversy). Sex is clearly not an option once you've become someone's meal. Placed in nearly any other environment, the queen angelfish, with its iridescent blue-green scales tipped with yellow edges and its brilliant yellow tail, would probably stand out, attracting predators. But the bright coloration of the angelfish is actually a form of camouflage; against the vivid backdrop of soft and hard Caribbean corals, the brilliant queen angelfish merely blends in. On the coral reef, a blandly colored angelfish would stand out like a three-piece gray suit in a biker bar.

But camouflage is only a beginning. Remember that these fishes live in shallow, brightly lit tropical waters, where colors remain vivid several feet down. Because of their environment, coral reef fishes have developed more ways than Matisse of using color. For example, several varieties of the petite butterflyfish have a single round, dark spot near their tail. Since many predators strike at their victim's eye, the false eye-spot may allow the butterflyfish an extra fraction of a second to escape an attack. The predator expects the victim to attempt an escape by swimming forward; but the butterflyfish darts off in the opposite direction. To further its deception, this same fish has a thick black line running vertically down its head, effectively masking its true eyes.

Biologists also theorize that some reef fishes have evolved striking coloration to let predators know they're unpalatable or even poisonous. Knowing a good idea when they see one, other fishes then mimic this same coloration to fool predators into avoiding *them* (a process known as Batesian mimicry). Of course, here I'm attributing intent to what is really an evolutionary process not governed by any decision made by individual fish. As in all such adaptations, natural selection is the only "intent." The adaptations found on the reef are so intricate and seemingly ingenious, however, that it's easy to fall into anthropomorphizing, even when you know better.

Take the astonishing case of the comet fish (*Calloplesiops altivelis*), a small, relatively rare fish found in the Indo-Pacific.[11] When threatened, the comet fish dives headfirst for any convenient hole in the reef, just like most reef fishes. But the comet fish goes a couple of steps further. First, it has the unusual behavior of leaving most of its body outside of the hole, with only its head tucked away. Any other fish would be gobbled up immediately. But rather than grabbing the exposed tail, potential predators beat a quick retreat. That's because the body of the comet fish has the same dark brown coloration with white spots as a common and highly aggressive species of moray eel, which lurks in similar holes, exposing only its head. To top off its impersonation, the comet fish expands its rear fin, exposing a strategically placed eye spot—making the fish nearly indistinguishable from the eel.

Predators also use color to capture prey. In eastern Indonesia I watched, mesmerized, as a long brown trumpetfish approached a yellow sea whip, changed its body color to match the sessile animal, and then hung in a head-down vertical position, mimicking the sea whip's arm while waiting for a meal to swim by. An extreme example of this ability to combine color and body shape is the anglerfish: a fish that fishes. Like the trumpetfish, the anglerfish's body is a study in camouflage. In texture and color it blends into the bottom reef community of sponges and algae-encrusted rock. Once in position, the anglerfish casts its lure, a modified dorsal spine to which is attached a small structure that looks remarkably like a tiny fish, and moves the bait in a circular motion only inches from its mouth. The piscivore (fish-eating fish) that lunges for the false fish ends up inside the anglerfish.[12]

Fishes use color in one of the more interesting interactions found on the reef: a process known as "cleaning symbiosis."[13] Many species of fish take part in this curious interaction, both as the cleaner and as the one being cleaned. A larger fish approaches some landmark, usually a large coral head. While hovering over the "cleaning station," the fish signals its desire to be cleaned, either by changing color (some species switch from a silvery shine to a dull red), by assuming an unusual position (some parrotfish stand "upright"), or by opening its mouth wide. These cues bring out the tiny cleaner fish, often gobies or wrasses, that live inside the coral head. The cleaner fish enters the mouth of the larger fish and begins the cleaning process, nibbling off parasites and dead tissue. When the job is done, the cleaner fish exits through the mouth or the gills and returns to its lair, while the larger fish returns to its normal color and position and swims off.

Perhaps the most remarkable part of this process is that if the two fish meet anywhere else, the larger fish will often eat the smaller one. Different rules apply at the cleaning station. A truce exists there, and you may sometimes see several large fish lined up awaiting this service. (A related fish, *Aspidontus taeniatus*, mimics the cleaner wrasse's coloration and behavior. But once inside the larger fish's mouth, this clever impostor takes a chunk out of its victim's cheek and darts out before the ambushed fish can react.)[14]

It's clear that the cleaner fish benefits from the relationship, but controversy swirls around the contention that such cleaning benefits the larger fish. It seems unlikely that larger fish would line up if they *didn't* benefit, but no reliable studies exist proving that this is the case, and some studies suggest that it may not be. "Cleaning" may not be a symbiotic process, but a parasitic one.[15]

A far more controversial subject is the role of coloration in fish reproduction. Some elements are uncontested, such as the fact that many sexually mature fishes have strikingly different coloration from juveniles of the same species. Colors are used in this case to signal related fishes that the fish in question belongs to the same species and is sexually mature. And, as with many other vertebrates (birds, for example), mature males and females of the same species frequently have different coloration.[16] But this is where things get a bit complicated, for sexual identity is a highly plastic concept among fishes of the coral reef.

Human foes of homosexual rights have adopted the glib bumper-sticker slogan "God Created Adam and Eve, Not Adam and Steve." Leaving aside the dubious accuracy of that statement as it applies to humans, on the coral reef things are a bit more complex.

A similar bumper sticker for reef fishes would read, "God Created Adam and Eve, Adam Who Becomes Eve, Eve Who Becomes Adam, and Adameve." And this is just the beginning of the confusion. Reef fishes have such diverse and inventive sex lives that they make pornography seem as bland and unimaginative as . . . well, as it really is.

In understanding the sex lives of fishes, it's best to throw away most of our human-based assumptions about sex. There are reef fishes that are born either male or female and remain so their entire lives, a condition known as gonochorism.[17] But for most of the fishes found on the coral reef, such fixed sexuality is downright kinky. It figures, then, that monogamy, too, is highly unusual.[18] Even copulation, the insertion of the male sexual organ into the female, is nearly unheard of.[19]

The primary mode of reproduction for reef fishes, as with fishes in general, is spawning, in which females release unfertilized eggs that are then fertilized by males.[20] The variations on this theme alone are staggering. Females may release enormous clouds of eggs into the water column (where most of them will perish), or she may lay fewer of them in protected nooks and crannies of the reef bottom, increasing the individual survival rate. Couples may pair off, with the male fertilizing the eggs of just one female. Or huge spawning aggregations may fill the waters with tens of thousands of fishes, releasing millions of eggs and sperm. A dominant male may have a harem of females with whom to mate. Fertilized eggs may be left to their fate, or cared for (brooded) by either the male or the female. The male seahorse, for example, broods his progeny in an abdominal pouch, carrying them around as a female kangaroo carries her offspring. Courtship rituals may be highly elaborate, with both fishes changing color and swimming in ritualized courtship dances—or nonexistent. Spawning may be influenced by tides, lunar cycles, seasons of the year, the specific terrain, or nothing in particular.

If such differences were simply interspecific, varying only among fishes of different species, that would be one thing. But on the coral reef many social behaviors, including the reproductive modes discussed above, vary among individuals of the same species. Ecologists explain this variability as a result of the complexity of the reef itself, with its micro-scale habitats separated by only feet, sometimes by inches. The most successful fishes in this complex ecosystem are those that can take advantage of this enormous diversity.[21]

The matter of sexual plasticity, of changing from one sex to another, has been studied in some detail over the past two decades with surprising results. It turns out that most families of reef fishes are sequential hermaphrodites, meaning that individuals can change from one sex to another. The most common form of this behavior is for females to change into males.[22] One of the most abundant reef fishes, the wrasse, is probably the most studied example of this phenomenon.

The bluehead wrasse is a small fish, with mature individuals ranging anywhere from an inch and a half up to just six inches in length. These wrasses reproduce in a type of mating system called a *lek* (a term coined originally for the behavior of birds). Around noon each day, a mass of the largest, most colorful male wrasses congregate in a specific courtship area, each fish staking out a territory that he vigorously defends against intruders. These males have a distinctive coloration, called *terminal phase* (or simply TP). The rear two-thirds of the body is neon green to blue-green, and the head is a beautiful dusky blue. These two sections are separated by a black vertical bar, followed by a similar white bar, and another black bar.

Smaller and less colorful TP males, perhaps sensing that they cannot compete with these elegant fishes, set up shop not far from the main lekking area. Mixed in among these lesser TPs are another group of males with a very different color pattern, or, more accurately, several different patterns, all lumped together under the term *initial phase* (IP). Although only males are TP, both males and females are IP.

So here we have two swirling masses of fishes, the large TPs at center stage and the IP males and smaller TPs on the periphery.

Enter the female.

She fights her way through the outer masses of lesser suitors, like a beauty queen elbowing aside crowds of hopeful but homely boys, until she arrives at last at the turf staked out by her Prince Charming. She is so intent on mating with her chosen male that she will refuse the attentions of even slightly smaller suitors if the object of her attention is occupied. Because the largest TPs spawn several dozen times a day, and females just once, the female wrasses frequently have to wait their turn.[23]

The sex act itself appears passionate, if brief. The TP male performs a looping dance, rising and falling, until the female arrives on the scene. Then he swims horizontally above her a few times, his pectoral fins quivering with anticipatory excitement. The female joins her mate and the pair make a dash toward the surface, where they simultaneously release eggs and sperm.[24]

Frustrated male IPs have developed strategies to mate with unwilling females. Just as the chosen pair consummates their passion, one of these lesser males may flash through the cloud of gametes, releasing his own sperm. With unusual playfulness, biologists have termed this behavior "streaking."[25] An IP male might also sneak into the boudoir and nudge the waiting female so that she is stimulated into releasing her eggs, which are then fertilized by the intruder (a process aptly named "sneaking").[26] Or the IP males can act as a group, surrounding a lone female before she reaches her target, and repeatedly touching her until she releases her eggs.[27]

Here's where it gets really interesting. The large TP male, the quintessence of maleness for the bluehead wrasse, may have started life as an IP male and, like the ninety-pound weakling who signed up for the Charles Atlas body-building course, developed into a full-blooded TP.

Or *he* could have started as *she*. While the fact that bluehead wrasses were protogynous (female changing to male) has been known for some time, scientists have only recently begun to decipher the subtleties of the process. In 1989 and 1990, marine biologists Robert Warner and Stephen Swearer documented that in the wild (that is, on the reef) sex change for bluehead wrasses is socially controlled.[28]

On separate Caribbean reefs, the biologists removed the largest, most brightly colored TPs. The results were immediate: "Several of the largest females present began to exhibit aggression towards other large females and to court smaller females *within minutes* of the removal of the large males," reported the biologists [my emphasis].[29]

Later on the same day, these same IP females actually courted and spawned—as males—with females. Color change was not quite as immediate or as pronounced as behavior. On the first day, the heads of large IP females began to darken, but only during the mating period. By the third day, several individuals had begun a permanent transition to the TP coloration. The experiment only lasted twenty-eight days, and none of the formerly IP females had completed the change to TP coloration by that time. But several were well on their way, and the heads of many others had taken on a permanent bluish tinge. More important is the fact that some had developed mature sperm eight days into the experiment.[30] By the end of the experiment, many of these formerly IP females (regardless of their coloring) had become fully functional males.

This ability to change from female to male is found among many of the most common reef fishes, including all parrotfishes and many gobies, cods, groupers, snappers, and damselfishes.[31] The anemone fish is rare in that it can switch from male to female. Regardless of the direction of the change, however, once a new sexual identity is assumed, there is no turning back. This is a once-in-a-lifetime opportunity. There are exceptions to this rule, however. The hamlet, a common reef fish in the Caribbean, is under no such constraints. Hamlets are true hermaphrodites (as opposed to the serial hermaphrodites listed above), equipped with both male and female reproductive systems. A hamlet may mate first as a male, then as a female, and then as a male again, changing roles several times during the course of a single night.[32]

Whatever the means, the end is the same: a fertilized egg, that same ball of jelly—so ludicrously insubstantial and yet so full of promise—by which we humans have perpetuated ourselves through time. An individual female coral reef fish may produce between 10,000 and 1,000,000 eggs annually.[33] Not all these eggs are fertilized, and even among those that are, the mortality rate is strikingly high—perilously close to 100 percent.[34] But for millions of years, enough individuals have endured to produce new generations of fishes.

Until relatively recently, reef fishes were thought to live a rather provincial existence (their spectacular sex lives notwithstanding). Like the human residents of most nineteenth-century rural American communities, coral reef fishes were assumed to stick pretty close to home. That was in part because scientists were blinded by the idea that the coral reef was a closed ecosystem, a stable "island entire of it self." Added to that was the fact that adult coral reef fishes *are* sedentary, tending not to stray far from their own neighborhoods.[35]

But then biologists started paying more attention to the larvae of reef fish, the transitional stage between egg and adult. They found that, like teenagers everywhere, the larvae were driven by wanderlust. This tendency is so pronounced among reef-fish families, and can last for so long, that one author has suggested that "reef" fish are primarily creatures of the open water, with a "reproductive phase" on the reef.[36] That may be an overstatement. But it points to an important fact: reef fishes spend a considerable part of their life cycle off the reef. The distance traveled in this open-water larval phase ranges from a few meters to several hundred miles, depending on the species, ocean currents, and many other factors, known and unknown. And, as new discoveries often do, this one has prompted even more complex questions, such as, When larvae leave the reef, where do they go? Even more important, where do they end up? The more ecologists have probed larval travels, the more they have had to rethink fundamental notions about reef-fish populations.

In fact, the term *population* itself has been largely replaced by *metapopulation* when discussing many issues related to reef fishes. The difference between the two terms is critical to our understanding of reefs themselves.

Population refers to a group of individuals of the same species, sharing a particular habitat in which they interact regularly through time. "Through time" is vital to the idea of stability within a reef, for it implies continuity over generations. Imagine a written life history of fishes on one coral reef. According to the notions of stability and equilibrium that have dominated human understanding of reefs, such a book would resemble a section of the Hebrew Bible: "Bluehead Wrasse I begat Bluehead Wrasse II, who begat Bluehead Wrasse III" and so on.

The concept of metapopulation upsets this established order. It refers to a population of subpopulations, distant in space and time, yet genetically linked. Bluehead Wrasse I may still be the ancestor of Bluehead Wrasse XV, but the two may exist on distant reefs. The tribal integrity so important to the biblical lineages falls apart on the reef.

The significance of this fact was summed up by an Australian ecologist who concluded that, given this new information, "it has become popular to regard these populations [of reef fishes] as 'open non-equilibrial' systems in which gains from external sources . . . exert more control over abundance and community structure than the carrying capacity of the local environment."[37]

In other words, reefs are far from being the closed, stable systems they were long thought to be. With fishes as with corals themselves, populations are dynamic and unstable, connected to and dependent upon distant reefs.

Corals and fishes also interact with each other. As already noted, corals provide shelter for many varieties of fishes. There is also the predator-prey relationship, but this is a rather short story, for few fish actually depend on corals for nutrition. The butterflyfish, whose small mouth and bristlelike teeth are designed for munching on tender coral polyps, is an exception to this rule.[38] Parrotfish often appear to be eating corals, but more often they're really eating algae growing on an exposed section of coral skeleton. It is through this indirect path of eating algae that fish have their main effect on coral—and probably on the reef community as a whole.

If you went looking for algae on a healthy reef, you'd probably find very little. And yet it is there, growing at such a fast rate that shallow coral reefs are among the most productive ecosystems on the planet.[39] The key to this paradox is the presence of large numbers of herbivorous fish that are devouring these rapidly growing algae as fast as they can grow. Biologists actually counted the number of times these herbivores took a bite from a single square meter of reef on an average day; the number ranged from between 40,000 to 156,000 bites.[40] It is not surprising, then, that herbivores may make up 50 percent or more of the fish biomass of an average coral reef,[41] even though they tend to be much smaller than carnivorous fish.

A critical result of this herbivory is that it removes algae that would otherwise smother corals. Biologists have tested this theory by placing cages around corals, excluding fishes of all types. The corals growing inside the cages were gradually overgrown with fine filaments of algae, while corals growing in the open nearby remained healthy.[42] In recognition of their role in limiting algae growth, and allowing corals to flourish, herbivorous fishes have been dubbed the reef's "immune system."[43]

Most reef herbivores are constantly in motion in search of algae—the piscatory equivalent of human hunter-gatherers. But at least a few fish have taken up farming. This was discovered by chance in the early 1970s by a marine biologist named Vine, while taking photographs of fish off the coast of Sudan in the Red Sea.[44] To attract fishes, he moved algae-covered pieces of coral rubble to a sandy bottom. Immediately a variety of herbivores descended on the rock, stripping it of algae and making for great photographs. After he had repeated this procedure a few times, the lightbulb went on over Vine's head. If these fishes were so crazy about the algae when he set it down on the sand, why weren't they eating it where it grew, just a few feet away?

He swam over to the algae site and hovered awhile. Soon he noticed something very interesting.

"Each time a fish approached the green algal-covered rocks," Vine reported, "it was chased away by the aggressive displays of dark grey pomacentrids. . . ."

The common name for the fish described by Vine is the damselfish, a name that carries a deceptive connotation of fragility and helplessness. As Vine quickly found out, when he observed three-inch damselfishes attacking parrotfishes more than two feet long, damselfishes are small but they defend their territory with the ferocity of tiger sharks.

"Indeed," he wrote, "on several occasions while working inside their territories, I have been forced to control an instinct to back away from their excited and sometimes frightening advances."[45]

And these "frightening advances" were coming from a fish the size of his thumb!

Inside their staunchly defended territories, damselfishes cultivate a garden of luxuriant algae. These fishes not only scare away other herbivores, but actually "weed" their algal patch, removing less desirable species so that their preferred crop will flourish.[46]

Those large parrotfishes chased off by the damselfishes play an equally important role in maintaining the local seascape of the reef and surrounding areas. As one of the many herbivores, parrotfishes help to keep algae in check. They also have the unique ability to bite off chunks of rock, using their eponymous parrot-like "beaks," and grind up the stone with the "pharyngeal mill" at the back of their throats.

When living in Key West, I used to delight in swimming among parrotfishes. Their beaks give them a smiling, clownlike appearance, and their brilliant colors are straight out of the circus. Their habit of eating rocks could be a sideshow attraction. But there are important ecological consequences to this behavior. What enters the parrotfish as rock, exits as sand—as it turns out, a surprising amount of sand. One study found that parrotfishes added over three hundred pounds of sand a year for every square meter of coral reef.[47] True, that was partly due to an abundance of parrotfishes at that particular reef, but the point remains valid: these gaudy fishes create a phenomenal amount of sand. That's one reason why the visibility on reefs is generally higher in the early mornings, before parrotfishes have begun their rock-munching. Some of this sand falls onto corals, which shed it by producing globs of mucus. But the bulk of the sand is carried away from the reef crest by incoming waves, where it mixes with sand produced by the countless other sources of reef erosion, biological and mechanical. This sediment gradually settles out in the lagoon. The process is repeated thousands of time a day, day in and day out, for thousands of years, layer upon layer of sand being deposited. Eventually an island is created. And to that island come creatures as remarkable as the ones that produced it.

9

Neither Brethren nor Underlings

Even though his movements are generally slow, his hearing is poor, and he has little in the way of brain, the turtle can be called one of the most successful animal stories in the world.

—Jack Rudloe, *Time of the Turtle* (New York: Knopf, 1979)

Dawn arrives sullen and gray on Heron Island, a high-latitude coda to the Great Barrier Reef. A damp chill hangs in the air. Far out at sea, rain is falling and the sky overhead looks bruised and raw.

I trudge along the shoreline, not expecting to run into a soul at this hour. The other guests are safely cocooned in their hotel beds, windows closed against the harrowing shrieks of the mutton birds, which continued their bizarre and plaintive cries throughout most of the night. Even the noody terns, jammed onto this small island by the hundreds of thousands, are still asleep in their thrown-together nests of grass and guano. Like cartoon characters, they chatter and squawk endlessly during the day, flitting about, crashing into trees, people, each other. Their nests fall apart, spilling eggs onto the ground, and then they rush about building new nests in the same slipshod manner and with the same disastrous results. Darwin called them "silly little creatures,"[1] and, seeing them at work, it's hard to argue with that assessment. But for now, thankfully, they are asleep.

I round a bend at the island's eastern tip, a place called Shark Bay. My eyes are cast down, my mind as fogbound as the sky, idiotically focused on the *skiff-skiff-skiff* sound my sneakered feet make on the wet sand.

Suddenly there she is, not ten yards in front of me.

I, too, become a cartoon character: my eyes bug out, my jaw drops, my heart begins to pound. I have never seen anything like her.

She is the size, shape, and color of a large boulder. Her shell alone is over a yard long, remarkably smooth and flecked with small patches of barnacles and a few threads of green algae. For once the common name lacks the grace of the scientific one. "Green sea turtle" is blandly descriptive. *Chelonia mydas* fairly sings.

She is exhausted, her great snout buried in the sand. As I approach, she lifts her massive head with great effort, but her eyes are unfocused and cloudy. My gaze follows her tracks up the beach, where they disappear over a small rise. Somewhere in there, at a place above the high-water line, she laid her eggs during the night. Approximately a hundred of the Ping-Pong-ball shaped eggs now lie buried beneath the sand, where they will incubate for several weeks. Her job accomplished, she is trying to return to the sea. The tide has gone out since she came ashore, and a low wall of dark limestone now blocks her passage. She must either pull her considerable bulk—well over two hundred pounds—over those rocks, or trace a path parallel to the water to where the obstruction ends and the beach offers clear passage to the water.

I sit at a distance and watch as she considers her options. A few people walk by. She doesn't move. She reminds me of my own wife after she delivered our own hatchling of one, lying dazed and oblivious of everything except her own fatigue.

A titanium tag with the characters "T31271" stamped into it is mounted on the turtle's right front flipper. Much later, through a series of transpacific faxes and E-mail queries, I learn that she was tagged while nesting on this same beach in December 1987. That year she laid at least three clutches of eggs, at intervals of a couple of weeks.[2] She has probably returned to Heron Island to nest once or twice between 1987 and this current visit.[3] Scientists believe that *Chelonia mydas* doesn't reach sexual maturity until between thirty-five and forty years old. This one was estimated to be over forty when she was tagged, which makes her over fifty now. If scientific estimates are right, she is just beginning her reproductive life. Although she will migrate to feeding grounds perhaps more than a thousand miles away, she will return to this same beach,[4] laying several clutches a season, every two or three years for another half-century.[5]

Finally I leave her to her attempts to regain the sea, and I head back down the beach. Now fully awake and invigorated from the encounter, I gaze out to sea, where the sun is beginning to break through the clouds. A flicker of movement in the water captures my attention. As massive and silent as rocks, nearly a dozen green sea turtles bob in the dappled shallows of the lagoon.

❄

"After you see them lay eggs, they get into your blood," says Leigh Slater, two days later, as we sit outside basking in the warm sunlight at the Heron Island National Park station, a few hundred feet from where T31271 buried her clutch. It is 4:00 P.M. and Leigh is bleary-eyed, still awakening after a long night spent circling the island conducting a turtle survey, recording and tagging nesters. She is a recent graduate of the University of Queensland. Like recent college graduates around the globe, she's figuring out what to do next with her life. While many young people in her situation take jobs in video stores or wait tables until they decide what path to take, Leigh has what must be one of the most interesting summer jobs on the planet.

"Yeah," she admits with a smile, pushing back a tangle of dark hair, "I'm pretty lucky."

Heron Island is a unique place in many ways. The forty-acre island sits astride the Tropic of Capricorn. Technically the northern edge is tropical while the southern part isn't. In addition to the National Park and a resort that caters to scuba divers, Heron Island is the home of Australia's largest university marine station (which is in a different time zone from the resort, only a short walk away). The island itself may be small, but the reef that created it encompasses some thirteen square miles. Add to this the many hundreds of acres of sea-grass beds, and you have a collection of habitats that is enormous, in both size and diversity of life. Besides the hundreds of varieties of corals and the thousand species of fishes and the possibly millions of invertebrates that make their home here, Heron Island is also the primary nesting area for green sea turtles in the southern half of the world's largest barrier reef.[6]

Sea turtles like *Chelonia mydas* move regularly between coral reefs and sea-grass meadows, a living link between the two ecosystems. What do they know of or care about the boundaries we humans draw? Jeremy Jackson, a highly respected biologist with the Smithsonian Tropical Research Institute in Panama, has argued that these megavertebrates have been keystone species for both—that is, the most important ecological resident in each ecosystem, analogous to elephants or wildebeeste on the African plain.[7]

For all their importance, however, sea turtles remain largely an enigma to biologists. Even many of the basic questions about their life history remain unanswered. The most authoritative scientific book on the subject, *The Biology of Sea Turtles*, is filled with statements such as "Absolutely nothing is known about the impact of factors such as aging, on reproduction,"[8] and "Surprising gaps remain in our knowledge of feeding habits."[9]

Steve Grenard, a naturalist and author of several books on reptiles, puts it even more plainly. "There is so much we don't know about sea turtles," he says, "it's unbelievable."[10]

An example of our ignorance is the period once referred to as the "lost year." This is the stage between the time the tiny hatchlings enter the water and when they reappear as large juveniles in a near-shore feeding habitat. Virtually nothing is known about this interval. A recent breakthrough in our understanding of this period is really an admission of greater ignorance. The gap is now believed to encompass up to ten years, not one. The "lost year" has become the "lost decade."[11]

"The thing is," Leigh Slater explains, when I continue to ask questions for which she has no answers, "we only have twenty-four years of data—and they have such a long life span."

Their life span is estimated to be more than one hundred years, but just how *much* more is another blank spot.

Scientific knowledge grows like the reef itself, through accretion and erosion. Data sets accumulate, giving rise to new theories that are attacked and defended. In time, many of these theories prove flawed. They crumble and fade, remembered, if at all, as quaint relics of an earlier age. But other theories stand up to scrutiny. Over time, they gain acceptance and a certain legitimacy—a standing that is temporary, for theory is always at the mercy of new data, new discoveries, new ways of seeing and making sense of the world. Science itself is the process of weaving disparate strands of information into a braid we call knowledge, a plait that is ceaselessly unwoven and rewoven.

The program Leigh is working for is one of many strands in this scientific effort, and so I'm delighted when she agrees to let me come along that night as she conducts her next survey.

"How will I find you?"

"Just start walking around the island when it gets dark," she says. "I'll be the one without flippers."

Well after sundown, when darkness has reclaimed the island, and the constellations of the Southern Hemisphere provide the only illumination, I head down to the beach and follow the shoreline, walking slowly clockwise around the island.

Green turtles nest only at night. As reptiles, they lack the internal mechanisms to regulate body temperature. If they had to come ashore to nest during the day, their massive, dark-colored shells would collect solar energy and they'd cook. Not long ago it was believed that sea turtles waited for the full moon to nest, but that theory didn't hold up. People only *noticed* them nesting when the full moon provided enough light. Nesting females are sensitive to light in a different sense. If there is a bright light on or near their nesting beach, they will avoid coming ashore entirely, or return to the sea without laying their eggs. For this reason I've been warned not to use a flashlight. I'm afraid I'll either miss Leigh in the darkness or, worse, trip over a turtle on her way to nesting and send her crawling back into the ocean. Fortunately, neither happens. I find Leigh at Shark Bay, not far from where I first saw T31271.

She's sorting out turtle tracks, muttering to herself.

When she spots me, she nods a perfunctory greeting and points to a trail in the sand.

"That's an 'up' track," she says, indicating something about the trail that I can't discern. (By starlight, I can barely make out the tracks in the first place.)

"That means she's got to still be up here someplace. Come on."

We climb up into the bush and immediately find the turtle, which is still moving about looking for a suitable nesting spot.

"Hello! There you are, girl," Leigh clucks. "Oh, this isn't a good place—too many roots, and the hill is too steep!"

She pulls out the tagging instrument, which resembles a staple gun, from a canvas equipment bag slung over her shoulder. She briefly switches on the dim light mounted on her yellow hard hat. She tags the turtle, riveting a metal marker bearing the inscription "T95801" to its flipper. "It doesn't hurt them," she assures me when I unconsciously mutter "Ouch!" as the tag is put in place. She hurries back down the beach and, with her heel, draws a line through the track to indicate that it has been recorded.

"Listen," she says, and holds up a hand.

At first all I hear is the sibilant sound of waves striking the shore. Then something else emerges: the faint, dry, rasping of sand being scattered.

"Let's go."

She strides toward the sound and then quietly approaches from the rear. I keep a few feet behind her. Because turtles have poor hearing on land, we are able to get within a few yards of her.

"She's just beginning to dig her body pit," whispers Leigh.

We watch for a moment as the giant turtle uses her powerful front flippers to clear the top layer of sand from the surface. When she's finished, she'll pull herself into this slight depression and begin digging an egg chamber, using her hind flippers.

"We'll come back later," says Leigh, and heads off into the darkness.

A few yards on, Leigh spots what she's been looking for.

"Oh, good, she's laying!" she says, her voice full of excitement.

Sure enough, this turtle rests in her body pit. She has already excavated a deep, flask-shaped egg chamber and has begun laying her small round eggs in it.

From a nearby clump of bushes comes the sound of more digging.

"Stay with her," Leigh whispers. "I'm going to check on the one over there."

Turtles enter a sort of trance when they begin laying eggs, and they are oblivious of most movement or sound. Still, I try to remain quiet so that I don't spook her. Besides, the scene seems to call for a respectful silence, as moments of birth and death always do. Kneeling on the sand directly behind her, I pull out my flashlight and carefully direct the beam into the egg chamber. Surrounded by darkness and framed by her hind flippers, the spectacle of the chamber slowly filling with soft and glistening eggs, each with its own minute, pulsing embryo, is like peering into an ancient temple where a sacred ritual is being enacted.

The process unfolding in the egg chamber is truly an ancient one. Sea turtles have been coming ashore to lay their eggs for millions of years, and their lineage stretches back even further, to more than 200 million years ago, when their terrestrial ancestors shared the earth with those other famous reptiles, the dinosaurs.[12] Why they first took to the water isn't known, although they were just one of many kinds of reptilians that did. For 100 million years their bodies adapted to the sea. Legs evolved into flippers. Bodies became streamlined for swimming, internal organs more suited to the ocean than to dry land. One innovation is the ability of these air-breathing creatures to take in enough oxygen in a couple of seconds to last them several hours underwater.[13] They also developed enormous salt glands, located behind their eyes, that allowed them to eliminate the salt they took in while living in the sea.[14] These strands of salty mucus coming from sea turtles' eyes led some observers to believe that nesting mothers wept as they gave birth.

No one knows how sea turtles survived the Cretaceous "crash," some 65 million years ago, when dinosaurs, including the seagoing ichthyosaurs and plesiosaurs, went extinct. A marine biologist I asked said it was likely that their hard shells played an important part in the sea turtles' evolutionary success—what she referred to as their "armored-tank strategy."[15] However they did it, marine turtles have managed to stay in business for 100 million years, continuously adapting to changes as needed.

Just as the egg chamber is nearly filled, Leigh returns.

"You about done, girl?" she asks pleasantly, as she quietly and efficiently goes about her data collection. She measures and records the size of the shell—just over a yard long—and draws a blood sample for measuring hormone levels. To obtain the blood, she gently pushes the turtle's large head downward, relaxing its powerful neck muscles and exposing a sinus cavity into which she inserts the needle, and pulls back on the plunger. As the syringe slowly fills with blood, Leigh croons, "Good girl!" over and over, although the turtle seems unaware of it all as her last eggs tumble into the chamber.

The turtle buries her eggs with the moist sand she excavated earlier, and then turns to head back to the sea.

"Wait, girl," Leigh says, a hint of alarm creeping into her voice as she fusses over the tagging device.

"Stop her!" Leigh commands as she hurriedly prepares the tag.

I haven't the faintest idea how to do this.

"What do I do?"

"Get over her and hold on to her shell."

Sounds simple enough. On the principle that pretending you know what you're dong will often see you through new and difficult situations, I confidently throw one leg over the sea turtle's enormous back, and with both feet planted firmly in the sand, I grasp her shell, each hand a few inches behind her head.

She doesn't even slow down.

Not that she's moving fast. The problem is that she is moving inexorably toward the water, dragging me with her like a crêpe-paper streamer attached to the bumper of a slow-moving truck.

Luckily, Leigh gets the tagger ready in time and allows me to dismount before the turtle can drag me indecorously into the lagoon.

Leigh expertly steps behind the left front flipper, immobilizing it with her foot, and hops on, murmuring, "Good girl" (although now through gritted teeth). She resembles a cowgirl riding a skittish horse, albeit in slow motion. The free right front flipper continues to pull at the sand, and the two of them turn a slow circle as Leigh finally manages to get the tag in place.

She climbs off, and we watch the turtle continue back into the sea.

"God, I hate that part," she confesses. "I'm just not strong enough."

Still, she gamely repeats this process, sometimes with ease, other times with more difficulty, around the island. There are so many turtles nesting simultaneously that the sound of digging seems to come from all directions. Leigh runs here and there, and I am essentially useless. It is like a large day-care center filled with rambunctious children and only one adult worker. This is a good nesting season—a record one, in fact—with more than two thousand females arriving during the short period.[16] Fortunately, the entire nesting process takes two or three hours,[17] so once Leigh spots a turtle emerging from the water, she knows she has some time to collect her data.

What seems fairly frantic to me is actually quite tame compared with what will occur on this same beach in a couple of months when the eggs hatch, an event known—appropriately—as a "hatchling frenzy."[18] There is good reason for the frenzy: after they emerge from their protective eggs and dig their way to the surface, the young sea turtles are at their most vulnerable. They are smaller than a bar of hotel soap, and weigh less than an ounce.[19] The short stretch of sand between the nest and the sea becomes a deadly gauntlet for these tiny creatures. Only one in a thousand hatchlings makes it to adulthood. Of those that don't survive, many end their existence only minutes after emerging from the nest, gobbled up by hungry predators. The hatchling frenzy is one form of that common survival strategy in nature: produce so many offspring, all emerging at the same time, that predators are swamped with prey and unable to consume them all. The individual hatchling is expendable; it is the population and, through it, the species, that must survive.

Predators are legion. Depending on the nesting location, they can range from birds to foxes and wild dogs to raccoons to ants.[20] The hatchling's only hope is to make it to the water, out past the "wall of mouths"[21] that is the reef, and out to the open sea, where predators are fewer. Emerging under the cover of darkness at night (which they deduce from the falling temperature of the sand above them), the hatchlings find the sea by heading toward the relative light—where stars are reflected in the waves.[22] When they hit the water, they dive into the undertow, rising again when they are well clear of the beach. Then they swim like crazy away from the island and reef predators for a full twenty-four hours before resting. At first they use incoming waves to orient themselves. But once clear of the island and reef, they can no longer depend on wave patterns. At that point the hatchlings switch over to some other system to orient themselves.

For decades, scientists have been mystified by sea turtles' ability to migrate between distant sites. As with nearly everything else about sea turtles, more questions than answers remain about this subject. But there *is* enough evidence to suggest that these marine reptiles, like bees and migratory birds, have an exquisite "map sense" based on a reading of the earth's magnetic field.[23] Scientists using satellite telemetry recently found that leatherback turtles migrated from nesting sites in Costa Rica nearly two thousand miles southwest, in a corridor three hundred miles wide.[24] This ability is so highly developed in green sea turtles that individuals feeding in sea-grass beds off the Brazilian coast are able to return to nesting grounds on Ascension Island, a nine-by-six-mile speck in the middle of the South Atlantic, more than a thousand miles away. To put this feat another way, imagine being led blindfolded to an enormous putting green and, without being told where the cup is—and still blindfolded—repeatedly sinking fifty-foot putts.

It is assumed that the few hatchlings to survive the initial frenzy spend the next several years at sea, gently floating with sea currents and inhabiting tangled seaweed "rafts" as they continue to grow. This is the "lost decade" about which little is known. From bits of evidence gleaned over the years, these young pelagic green sea turtles appear to be omnivorous,[25] eating some seaweed, but probably preferring small invertebrates such as snails, jellyfish, and "comb jellies" (tiny, transparent relatives of jellyfish, mostly known for their spectacular bioluminescent displays of greenish-blue light when disturbed).[26]

A decade or so later, when these juveniles are at least dinner-plate size (too large to be eaten by anything but the largest sharks), they leave this life of drifting and carnivory in the open ocean to settle down in specific shallow-water feeding grounds (hundreds and sometimes more than a thousand miles away from their nesting grounds), becoming vegetarians as they feast on sea grasses and algae—although they are known to gulp down the occasional jellyfish or sponge when the urge for animal protein strikes.[27] The green turtle is so strongly associated with sea grasses that the common name for the dominant sea grass in the Caribbean, *Thalassia testudinum*, is "turtle grass." This linkage derives from the scientific name, for *testudinum* comes from the order Testudines, a grouping that includes all turtles.

For several decades, these older but still sexually immature juveniles graze sea-grass meadows and eat various algae on coral reefs. Along with the adults in the same location, these young turtles profoundly influence both ecosystems.[28] One way green turtles affect sea-grass beds is by establishing "grazing plots" in which they continually eat from the same plants, cropping them down close to their base. New growth has a higher protein content than do older leaves, so, by "regrazing," turtles improve their diet while simultaneously altering the nutrients available to the sea-grass ecosystem. The turtles then "export" nitrogen from the sea-grass beds to reefs and nearby areas (through feces and urine), altering those ecosystems as well.[29]

Green sea turtles can eat an enormous amount of sea grass. In fact, they have to: as cold-blooded reptiles they need to burn a lot of fuel to maintain their body temperature. Each mature female needs approximately one thousand square feet of turtle grass per year for her grazing plot[30]—an area slightly smaller than one side of a tennis court. In addition, green turtles eat large amounts of macroalgae growing on the reef itself. Given this diet, a group of a few thousand sea turtles would have a profound influence on sea-grass beds and nearby coral reefs.

Coming after years of mostly sedentary life, sexual maturity renews the green sea turtles' commitment to migration. Their wanderlust is exactly that: wandering prompted by lust. Both males and females leave their foraging grounds for distant courtship/mating areas near their nesting sites. These sites are close to, if not identical with, the beaches on which they hatched decades earlier.

An internal "map sense" may explain how turtles find a specific distant location, but it does nothing to account for how they know which is their natal (birth) beach. At first the explanation was found in the "social facilitation hypothesis." Basically, this theory amounts to a follow-the-leader-type exodus, where first-time breeders tag along after more experienced turtles. That seemed plausible. But then came DNA testing. Scientists observed that green turtles from Heron Island shared feeding grounds with others of that same species from Raine Island, in northern Australia. Yet, after analyzing the DNA in turtles nesting at each island, researchers discovered that each group was a distinct hereditary population. The two groups spent years together at the feeding ground, yet when it came time to mate, they separated, each group heading toward its own nesting territory.[31] This ruled out the "social facilitation hypothesis." However, marine biologists still don't know how green turtles are drawn to the region of their birth.

When sea turtles arrive at the courtship area, they begin what can only be described as a reptilian orgy, with males and females mating with many different partners over a period of several weeks. Even the French must stand humbled before the amorous ability of the sea turtle, for the act of copulation, which usually takes place on the water's surface but may occur deep underwater, lasts several hours. Captive pairs have been known to stay locked in the coital embrace for over ten hours.[32] One interesting feature of sea turtle reproductive anatomy is the fact that the green turtle's penis has twin openings, which allow it to deposit semen in each of the female's two oviducts during mating.[33] The semen from several males is then mixed together and stored until ovulation. Despite their generally placid reputation, "rough sex" is the norm for sea turtles during courtship and copulation. Males bite females, and bite each other as they fight over females, in both cases resulting in sometimes severe and even debilitating wounds.[34]

When the copulating period is over, the males make the return trip to the distant feeding grounds, while the females begin their nesting period.[35] This may last several months and usually results in several clutches of eggs. Then, she too heads back to the feeding ground, where she will rest and recover for two or three years before repeating the process.

I continue to follow Leigh, helping out as I can, while she continues to tag and record information on the nesting females. At one point she mentions that hatchlings from the nests we've observed so far will be mostly males.

It occurs to me that Leigh is pulling my leg.

"How do you know that?"

"Because we're on the north side of the island," she replies, as if that explains everything.

Now I'm sure she's joking. She must sense my doubts because she adds, "See, this is the cooler side. Less sunshine."

I still don't get it.

"Their sex isn't determined when the eggs are formed," she says. "It's only later that they become male or female. You see, it's the temperature that makes them one or the other. So these"—and she indicates the nests on the northern edge of Heron Island—"will mostly be males. On the southern side, where there's more sunshine, they'll be mostly females."

"Oh."

"Really."

"Okay."

"*Really.*"

"I believe you," I say, and make a mental note to check this out the first chance I get. (When I do finally pick up a copy of *The Biology of Sea Turtles*, I'm chagrined to learn that Leigh was absolutely right.)

She shakes her head and smiles.

"Come on," she says. "I hear another mother-to-be."

For another hour or so I follow along, watching by starlight as these amazing creatures emerge from the lagoon, one after another, lumber up the beach, dig their nests, deposit their eggs, and then, just as laboriously, return to the water. In a few weeks, I know, they will all be gone, far out at sea, using a sense we neither possess nor understand to navigate the long voyage across the open ocean to their feeding grounds. I am reminded of a favorite passage by the American writer Henry Beston:

> For the animal shall not be measured by man. In a world older and more complete than ours they move finished and complete, gifted with extensions of the senses we have lost or never attained, living by voices we shall never hear. They are not

> brethren, they are not underlings; they are other nations, caught with ourselves in the net of life and time, fellow prisoners of the splendour and travail of the earth.[36]

Maybe it's because sea turtles remain such a mystery (and who is immune to the allure of the unknown?). Maybe it's because they are ancient and worthy of respect—if not veneration—on that count alone. And maybe it's the knowledge that, despite all the millennia that sea turtles have existed on the planet, owing to the activities of my species, *Chelonia mydas* may not be around in another hundred years. Some of their close relatives already teeter on the edge of extinction. (Kemp's ridley sea turtle is a good, if extreme, example. In 1947, some 40,000 female Kemp's ridleys arrived at their primary nesting site, a beach in Mexico. In 1966 only 1,300 showed up. Today, about 600 remain.[37])

Whatever the reason, and against my will (for I am suspect of my own sentimentality in regards to animals), as they slip into the sea, their enormous glistening backs disappearing beneath the water, I hear myself whisper to the departing turtles a word I have never uttered in seriousness: "Godspeed." The word is as archaic, or so it seems to me in human terms, as the sea turtle herself.

I'm certain Leigh didn't hear me, but perhaps she reads minds, for as if to temper my good wishes with the fire of reality, she follows my gaze and says simply, "She'll probably end up being eaten in Indonesia."

PART TWO

Human/Nature

What we must do is encourage a *sea change* in attitude, one that acknowledges that we are a part of the living world, not apart from it.

—Sylvia Earle

10

The Jakarta Scenario

It is the city which should be judged though we, its children, must pay the price.

—Lawrence Durrell, *Justine*

The pretty young woman standing behind the desk at the Ancol Marina is pleasant but insistent.

"One million rupiah," she repeats, smoothing her official blue skirt. No, she says, there is no cheaper way to Nyamuk Besar, a tiny island barely three miles out into Jakarta Bay.

She continues to smile in a manner that is meant either to cushion the blow of the exorbitant price (assuming we're stupid enough to pay it) or to invite further negotiations (in case we're not quite *that* stupid). Bargaining is a serious artform here on Java.

Stefano Fazi, an Italian marine biologist with UNESCO who is accompanying me for the day, whispers in a bewildered tone, "One million rupiah? That's a lot of money, no?"

It certainly is—roughly $450 for a boat ride that couldn't take more than half an hour each way. It seems even more expensive considering that Nyamuk Besar may not exist. A respected Indonesian scientist had informed me with grave certainty only the day before that the small island disappeared years ago, eroded into oblivion.

"It is gone," he had said, pointing sadly to the field of blue on the map spread out on his desktop. "All gone. Covered by water."

I silently pray he was mistaken as I attempt to match the young woman's smile with one equally amiable and just as determined.

"No," I say, shaking my head. "We can't pay that much."

She shrugs and bites the inside of her cheek. The three of us stand in silence for a minute. Finally she says, "Wait here," and walks off, her heels clicking her retreat on the wooden floor.

In her absence, Stefano unleashes his full shock.

"We can't pay that!" he fumes. "That's more than it costs for a weekend at a fancy resort in the Kepulauan Seribu, including the ferry!"

Of course he is right. The Kepulauan Seribu, or Thousand Islands, is an archipelago that begins off the western tip of the bay and stretches north into the Java Sea for fifty miles. The name Kepulauan Seribu is as inflated as the initial price quoted by the Ancol Marina hostess. In truth, there are only slightly more than one hundred islands in the chain. Many of them are privately owned, weekend getaways for Jakarta's chummy elite. On others there are beautiful hotels, the kind of places where tall, cool drinks appear at your elbow as if by magic as you recline in indolent luxury on white sand beaches.

Stefano has been posted to Jakarta for just three months. He has seen little of the city outside of the corridor that connects his apartment to his office in the UN building (which is in a swanky section of Jalan Thamrin, directly across the boulevard from Indonesia's first McDonald's and its only Hard Rock Café). When we met in his office the day before, I had mentioned that I planned to hire a boat and visit Nyamuk Besar. Stefano nearly leapt with excitement.

"I can go with you?"

It was not clear from the syntax whether this was a question, but either way I figured it would be nice to have him along. Especially good to have company if it turns out that the other scientist was right and Nyamuk Besar is submerged beneath the waters of Jakarta Bay.

The woman finally returns. Her heels are still clicking, but she has turned down her smile a few notches. Not a good sign.

"One half million rupiah," she says and immediately looks away, flipping her shoulder-length hair at us in the process. I take this as a signal that bargaining time is over. We have reached the "take it or leave it" phase.

I look at Stefano. He wears the expression of a man who has just bitten into a lemon.

"Nothing cheaper?" I ask her.

She stares off in another direction and chews the inside of her cheek again.

Stefano is prepared to call it a day. But I haven't come all the way to Jakarta just to have my plans to visit Nyamuk Besar evaporate at the first hurdle (or the second hurdle, if you count the possibility that the island itself has evaporated). I had read about the mysterious and exotic Nyamuk Besar weeks earlier, back home in Iowa, and was instantly captivated with the idea of visiting it. Granted, there are few places on earth that do not sound mysterious and exotic from a base camp in Iowa. But still, back in the 1930s a straitlaced young Dutch geologist named Johannes Umbgrove was so moved by what he had seen in the crystal-clear waters surrounding the island that he breached scientific protocol and waxed poetic in a scholarly paper:

> The unrivalled splendour and wealth of forms and the delicate tints of the coral structures, the brilliant colours of fishes, clams, sea anemones, worms, crabs, star fishes and the whole rest of reef animals are so attractive and interesting that it seems impossible to give an adequate description of such a profusion of serene and fascinating beauty.[1]

We thank the young woman and head out to the docks. In the shade of a thatched bench, a scrum of men of various ages sits passing the time and trying to stay cool in the heat of Jakarta, which, even by the ocean, is stultifying.

I say good morning in my best Berlitz.

"*Selamat pagi.*"

"*Pagi,*" they reply as a group.

I pass around a pack of Kretek cigarettes, the clove-and-nicotine-spiked medium of introduction among men in Indonesia.

Each man smiles in return and takes a cigarette. When the pack returns to me, I remove one of the aromatic, hand-rolled cigarettes and we all light up and puff away in silence for a moment.

Then I utter the sentence I had painstakingly cobbled together from my phrase book in my hotel room the night before, and had practiced in my head during the long cab ride over to the Ancol Marina.

"*Saya ingin kapal ke Nyamuk Besar.*" I say it with authority, implying a false command of Bahasa Indonesian, the official language in a country of three hundred tongues.

The men laugh, and chatter between themselves.

"What did you say?" Stefano whispers.

"I thought I said I want to hire a boat to Nyamuk Besar."

"Good. Now tell them we can't pay a lot."

One young man, probably around Stefano's age, in his early twenties, stands up. He wears a serious look beneath his Chicago Bulls hat.

"Come with me," he says, and walks down the wooden pier. For the next half hour we follow him around the docks as he acts as our mediator. He relays our request to the captains of various vessels, ranging from a forty-foot luxury yacht to a tattered rubber dinghy complete with busted outboard motor and wooden paddles. After each exchange he returns with a price—always too high, but lower than the official Ancol Marina rate.

It is getting late, nearly noon, and we are hot. The air is stifling. The temperature is in the nineties and so is the humidity. Finally our guide finds a man who is willing to take us out to the island, stay there for twenty minutes, and bring us back, all for 200,000 rupiah, about forty dollars each. Still too high. But if we want to go, this is probably the best we'll do.

We shake hands on it and pay our intermediary for his efforts. In a few minutes our boat, a new twenty-five-footer with powerful twin Yamaha engines, is on its way. We thread our way slowly out of the marina and, at last, roar into Jakarta Bay. The salty spray is cool and pleasant on our faces.

The captain of the boat stands at the helm, silent and authoritative behind his sunglasses. The first mate is immediately friendly and chatty, like most Indonesians. The fact that we each know only a few words of the other's language is a minor impediment. We exchange names. His is Mohamed Hurun. When I tell him I am American, he smiles.

"Bill Cleen-ton" he says.

I nod and add, "and Muhammad Ali."

He cackles, doubled over with joyous laughter. Fifteen minutes later, Mohamed climbs nimbly onto the bow and starts to shout directions to the captain, pointing left or right as he spies rocks in the murky water. Several islands come into view as we head north at half speed. Mohamed points to the closest island and shouts, "Nyamuk Besar!"

For all his friendliness, I'm skeptical.

I think, We are just tourists to him, wealthy foreigners, with money for the taking like a ripe mango from a tree. Why not just deliver them to the closest island and tell them, "Here you go!" I am ashamed of my thoughts, but I know this is a possibility. From my backpack, I remove aerial photographs of Leiden, as the Dutch called Nyamuk Besar when Indonesia was their colony and Jakarta Bay was the Bay of Batavia. I study the pictures, and as we draw closer it is clear that this really is the right place, the tropical isle of Leiden that captivated Umbgrove some sixty years ago with its multiplicity of life.

The basic form of the island is the same thin spit of earth and trees, curling a bit to the northwest like a beckoning finger, edged by a lacework of sand and seaweed. The captain slows the engines to a low, rumbling crawl and circles the island. I lie on my stomach and gaze down into the water, looking for Umbgrove's "profusion of serene and fascinating beauty."

All I see is rock, lifeless and gray.

Where is the "splendour and wealth of forms and the delicate tints of the coral structures, the brilliant colours of fishes, clams, sea anemones, worms, crabs, star fishes and the whole rest of reef animals"?

There is no living coral at all. There are no fishes. The only "delicate tints" and "brilliant colours" in the water come from the multitude of plastic bags, chunks of Styrofoam, and liter soda bottles that float around the island. Oil slicks add an iridescent dash of color to the surface. Armies of the sea urchin *Diadema* rest in craggy rocks, built by the now-dead corals. This is not a good sign, however. It is, rather, an indication that much is wrong here. The predatory fishes that would have kept the *Diadema* population in check have likely been removed through overfishing. The urchins, with their long black spines, are a sign of ecosystem restructuring. *Diadema* eat algae, which bloom in nutrient-rich waters inhospitable to corals. A thriving population of *Diadema* suggests lots of algae, which in turn implies few corals.

As we circle the island I see only algae-covered rocks, *Diadema*, some coral rubble, and trash.

"All dead—very bad," says Mohamed. "This island finished."

Having completed the funereal circuit of Nyamuk Besar, the boat heads back south to Jakarta.

The devastation at Nyamuk Besar is shocking but it doesn't come as a surprise. And the problem isn't limited to this one island. Before leaving the States I had read a scientific paper that concluded "the coral reef communities in Jakarta Bay are examples *par excellence* of the terminal stage of coral reef eutrophication."[2] *Eutrophication* is the scientific term for what happens when nutrients enter naturally clear, nutrient-poor water, causing an explosion of algae.

There are a number of intermediate steps involved, but one can generalize with this simple equation: Nutrients in, corals—and nearly everything other than algae—out.

Nyamuk Besar is just one well-documented site in Jakarta Bay, a body of water that went from being a coral paradise to what can only be described as an algae-choked sewer in the span of a few decades. You now have to travel fifty miles from Jakarta to find the healthy reefs that not long ago existed in the bay itself.[3]

To get a sense of the enormity of this transformation, consider that when Umbgrove did the research he wrote about in 1939, he counted ninety-six species of hard coral. That's nearly a hundred species in an area that would fit comfortably into Lake Tahoe. In the entire Caribbean there are only about fifty species of hard corals.

Today only sixteen coral species are found in Jakarta Bay, and even that number gives an overly optimistic impression, for few colonies exist, and those that survive are sickly and probably dying. The same paper cited above referred to these once-luxuriant reefs as "functionally dead,"[4] adding that "today [1993] none of the coral reefs in Jakarta Bay can be considered as functional coral reef communities."

Another scientist put it more succinctly: "We've come up with a good word to describe these reefs," he told a colleague after returning from Jakarta. "Fucked."[5]

What has happened to Jakarta Bay? How did it go from coral paradise to coral graveyard in such a short time? And why should people outside of Indonesia care?

To answer these questions, we must turn to the city that gives its name to the bay.

Jakarta is a nightmare of a city. If, as scientists conclude, the corals in the bay are "examples *par excellence* of the terminal stage of coral reef eutrophication," then Jakarta is an example *par excellence* of the overwhelming problems facing modern megacities of the developing world.

Jakarta has been called "Asia's most overcrowded, underplanned, and least-visited capital."[6] The island of Java, on which Jakarta is located, is itself overcrowded, with 7 percent of Indonesia's land area and 60 percent of its total population.[7] As in other areas of the developing world, a massive exodus from rural areas has swelled Java's urban population over the past few decades. In 1970, one in six island residents lived in cities; today that ratio has changed to one in three.[8]

Jakarta is one of the best examples of this rural-to-urban trend and of its catastrophic consequences. The city sprawls from the mountains of the Bogor district, in the south, across a flat coastal plain to the Java Sea in the north, covering over 255 square miles. Even this large area isn't sufficient to accommodate the more than 10 million people who live there—at least not in a manner that could be considered sustainable.

The usual urban extremes of wealth and poverty are found in Jakarta, but here the rift separating the two widens to canyonlike dimensions. It is no exaggeration to say that there are two Jakartas. There is the "Golden Triangle," truly a gilded province of tinted-glass skyscrapers, luxury hotels with soaring atria, and Western-style shopping malls dotted with cool, gurgling fountains, where businessmen in hand-tailored suits from Hong Kong stride down tidy sidewalks, cell phones nestled into their silk pockets. The only sound disturbing this picture of tranquillity comes from the heavy equipment that is everywhere erecting more buildings, financed by some of the billions of foreign dollars that pour into Indonesia annually.[9] This is the showplace of the modern Indonesia. It is the handsome, smiling face that the government of President Suharto wants the world to see. That same government diverts foreign attention from its long-standing political and ethnic repression by pointing to the economic gains of the last twenty-five years. (At least a half-million people were slaughtered following the coup that brought Suharto to power in 1965–66.[10] The 1991 army massacre of 171 mourners at a funeral in Dili, on East Timor, was only the most egregious and recent human-rights violation in the government's continued repression of that former Portugese colony, annexed by Indonesia in 1976 after it was briefly independent.)

The economic gains are real enough, though probably overstated. According to the World Bank, the poverty rate for the nation as a whole has dropped from 60 percent to 15 percent over the past decades, and life expectancy at birth has jumped some twenty years.[11] The widespread malnutrition and even starvation that was the rule in the past has become a rarity, especially since the country became self-sufficient in rice production in 1985.[12]

But stray from the Golden Triangle and into the poor neighborhoods of Jakarta, especially in the outlying areas, and you begin to wonder about official statistics. This is the other Jakarta: noisy, squalid, overcrowded, and disease-ridden. Several neighborhoods have staggering population densities, with 130,000 people crammed into each square mile.[13] An estimated one-quarter of Jakarta residents are squatters, living illegally in huts of scrounged tin and rotting wood. This doesn't include the many completely homeless, the so-called *orang gelandangan*, or "people on the move," who sleep on the streets[14] and are probably not counted in official statistics.

Then there is the traffic. Jakarta does not *have* traffic jams; its essential nature *is* the traffic jam. That is largely because the number of motorized vehicles in the country has doubled in the past decade.[15] For hours at a time, several million engines idle throughout the city, pumping carbon monoxide and lead into the air as if this were their primary purpose (unleaded gasoline is not used). The vehicles housing these engines remain stationary, stuck bumper-to-bumper. Their occupants thumb through newspapers or read books or talk on cell phones. Occasionally they may look up and tap their horns in futile and tepid complaint. Then, suddenly, the swarm of cars, trucks and innumerable three-wheeled motorcycle-cabs called *bajaj* lurch forward, horns blaring, looking for openings that mostly do not exist, and weaving perilously between lanes when they find one. A few hundred yards on, movement once again comes to a halt, and the cycle begins anew.

The legions of *bajaj*, with their especially polluting two-stroke engines, owe their existence to a 1970s decree by President Suharto, who decided to modernize the city by ridding it of the traditional *becak*, the pedicabs found throughout Southeast Asia. All *becaks* were rounded up and tossed into Jakarta Bay. Not surprisingly Suharto's plan backfired. The *becak*s were instantly replaced by the noisy and polluting *bajaj*.

The fumes from all these engines, not to mention those from electrical generating plants, and from the many industrial facilities throughout the city, fill the hot and humid air. In the nearby city of Semarang, researchers analyzing air samples over a two-week period recorded five days on which the concentration of lead particles exceeded 10 micrograms per cubic meter—with a high reading of over 16.[16] Consider that the maximum safe level in the United States is currently set at 1.5 micrograms per cubic meter (the Environmental Protection Agency is currently suggesting that that level be revised downward to .5 micrograms per cubic meter).[17] There's little reason to think that Jakarta's air quality is superior to Semarang's, and so it's not surprising that respiratory problems account for over 12 percent of deaths in Jakarta.[18] Some of these heavy metals settle into Jakarta Bay, where they mix with other industrial chemicals and cause widespread havoc in the marine ecosystem, disrupting reproductive cycles and leaving fish prey to a variety of weakening or fatal diseases.[19] Mercury, a byproduct of Jakarta's industries, has been found in the bay in concentrations up to three times the maximum safe levels,[20] and reports of Minamata disease caused by this pollution are common among environmentalists and scientists in the area.

Even more deadly to corals than heavy metals is the introduction of nutrients into the water. In this, the two Jakartas, rich and poor, are indistinguishable, for whether human excrement is flushed down porcelain commodes in bathrooms of imported tiles, or whether it simply runs through grooves notched into the streets of the worst slums, it all ends up in the same place: Jakarta Bay—and nearly all of it is untreated. A decade ago, when Jakarta's population was significantly smaller than it is today, the city produced an estimated 7 million cubic feet of septic-tank discharge per day.[21] That's the equivalent of seventy-five Olympic-sized swimming pools of sewage flowing into the bay each day. In addition, one-third of Jakarta's solid waste—or 2,200 tons per day—ends up in the city's rivers and canals, and most of that also finds its way into the bay.[22] For humans, this fouling of waterways results in a chronic public health crisis (if a crisis can be called chronic), mainly owing to contamination by *E. coli* bacteria, which exist in Jakarta's canals at a level some five thousand times higher than the standard for drinking water.[23] While few drink this polluted water directly, it frequently contaminates drinking supplies, and a significant number of people bathe in Jakarta's canals where they are exposed to the bacteria.

This nutrient-rich muck could have doomed the coral reefs in Jakarta Bay by itself—especially when you add in the agricultural nutrients coming from upstream. Between 1964 and 1992, the amount of phosphates in Jakarta Bay doubled; in the same period, the amount of nitrates increased more than fivefold.[24] As a result, macroalgae blooms are now regular features throughout the bay, choking corals and asphyxiating other marine life.[25]

Another factor implicated in the death of Jakarta Bay is sedimentation, with large quantities of silt entering the bay from Jakarta's three main rivers. Human activities upstream, including mining, logging, and agriculture, add a tremendous amount of sediments to the water flowing into the bay, fine particles that prevent sunlight from reaching corals and their photosynthetic algae.[26] Along with these sediments come pesticides, such as DDT.[27] Much of the sediment flowing from these rivers had previously been trapped by coastal mangrove forests, but those have been cut down over the last few decades, allowing the silt and pollution to flow unimpeded into the bay.[28]

Even the plastic bags drifting in the waters around Nyamuk Besar are not just an aethestic affront; they, too, play a role in the destruction of Jakarta Bay. What one scientist has called a "blanket of plastic over the inner Thousand Islands sea bottom"[29] likely prevents coral larvae from settling and establishing new colonies—assuming corals could survive and reproduce, given the other environmental assaults.

It's not just what goes into the bay that has rendered it functionally dead, but what humans have thoughtlessly removed from it. The calcium skeleton laid down by corals over several thousand years has been mined from Jakarta Bay since at least the mid-1800s, for use in construction. The pernicious effects of this practice were already evident in the 1940s, when Umbgrove compared a map of the bay drawn in 1753 with a contemporary Dutch naval chart. He found that "two centuries ago the small sand cay [island] of Schiedam in the Bay of Batavia was a rather large low-wooded island, and at that time the now-submerged reef 'Vader Smit' was the largest cay in the bay!"[30]

In the 1920s, coral mining was a big business in the region, with up to a million cubic feet of coral extracted annually ("in a good year," as one report puts it).[31] Unfortunately, this devastating practice is not a relic of the past. The island of Air Kecil was literally hacked apart, cut into blocks of coral (including living colonies) in the early 1980s to build the runway at Jakarta's Sukarno-Hatta International Airport.[32] In November 1996 a scientist, visiting the central Javanese island of Karimunjawa, found that three-quarters of new construction still uses coral "as the primary or exclusive building material."[33] Ironically, and most disturbingly, this destruction takes place on the main island of the Karimunjawa National Marine Park.

In addition to the mining of coral blocks, coral sand and gravel are still dredged from Jakarta Bay at an ever-increasing rate, despite laws prohibiting the practice. The dredging not only hurts corals by increasing turbidity, but, as Umbgrove pointed out, leads to the extinction of entire islands, which, without their buffering reefs, are slowly washed away by waves and currents. Nyamuk Besar, Umbgrove's beloved Leiden, which grows smaller by the year, appears headed for that fate.

Scientists have only recently begun to study the negative effects of overfishing on coral reefs. It appears likely, however, that increased human population has led to increased pressure on fishing stocks in Jakarta Bay—which, in turn, has disrupted the bay's complex food web, which includes corals. The signs of this process are unmistakable from the air as you fly over Jakarta Bay approaching or leaving Sukarno-Hatta International Airport. In the murky waters below are hundreds of what appear to be giant floating crates. These are bamboo frames for lift nets, called *bagans*, used by fishermen to haul their catches from the bay. But more and more fishermen are competing for fewer and fewer fishes. For the entire Thousand Islands fishery, the total catch is about one-tenth of what it was two decades ago.[34] And when too many herbivorous fishes are removed from the waters, algae (already given a boost by the nutrient pollution) cannot be kept in check, and a host of marine animals, including corals, suffer.

Human waste. Heavy metal pollution. Sedimentation. Coral mining and dredging. Overfishing. What killed Jakarta Bay? That's like asking who killed a man executed by a firing squad. In the case of Jakarta Bay, any one of the environmental insults could have conceivably done the job by itself. Taken together, its death was inevitable.

No one would argue that what has happened to Jakarta Bay is anything short of a tragedy. A rich and flourishing, biologically diverse ecosystem has, in an astonishingly short time, been wiped out by human activity. But that brings us to our second question: Is the destruction of Jakarta Bay important to people outside of Indonesia?

I believe it is, for two reasons. First, as the center of planetary marine biodiversity, Indonesia holds a unique position on Earth.

What is lost to Indonesia is lost to the entire world—and we can ill afford such a decline in biodiversity. Jakarta's problems, largely based on population growth, are spreading to other areas of Indonesia, where ten cities now have populations in excess of one million and show no signs of slowing their growth.[35]

But there is an even more compelling reason to learn the lesson of Jakarta Bay: it may be a harbinger of a worldwide decline or, at worst, of a near-total collapse of coral reefs. Such a collapse would be catastrophic beyond our imagining. Without living reefs to match ever-rising sea levels, the coastlines of many tropical countries would be slowly drowned, and exposed to greater damage from tidal waves caused by earthquakes and the surges associated with hurricanes. Some low island nations in the Pacific would disappear altogether—just like some of the islands in Jakarta Bay. Without the reefs, entire fisheries would collapse, causing malnutrition if not outright starvation. The ensuing chaos would set off a chain of events, the end of which is impossible to predict.

"What you're really talking about is *political* instability," says Mike Risk, a well-respected reef scientist who has studied the situation in Indonesia. "As the reefs go belly-up, we'll see five new countries [in Indonesia]."[36]

The environmental/economic/political stakes go beyond coral reefs, which have been called the "miner's canary" for terrestrial ecosystems. During the long span of geologic time, whenever there has been a mass global extinction, it has been preceded by the collapse of reefs.[37]

There is evidence that the Jakarta scenario is, in fact, being played out globally. Of course, there are variations on this theme. The causes of coral destruction vary depending on the location. And few areas have reached the state of utter devastation found in Jakarta Bay. It is also true that many reefs appear to be thriving. But scientists are finding early evidence of decline all around the globe, from the Caribbean to the Red Sea to the Great Barrier Reef. And much of the destruction appears to be caused by human activities. Exactly how much isn't known. A recent survey of the status of coral reefs in Southeast Asia (home to one-third of the world's reefs) showed startling declines.[38] In Indonesia, only 2.6 percent of reefs are estimated to be in excellent condition, while three-fourths are in the fair-to-poor category. The figures from the Philippines are even worse, with 1.3 percent of reefs rated excellent, and over 90 percent considered fair-to-poor. Thailand is doing slightly better (with almost 17 percent of reefs found to be excellent, and 41 percent in the fair-to-poor group), but is still a cause for a concern. Even more alarming is the fact that the overwhelming majority of this destruction has taken place in just the last fifty years.[39]

The growth of megacities like Jakarta, those containing 10 million people or more, should in itself be seen as a warning light, for experience shows that coral reefs have a hard time coexisting with human population centers. And until we have a handle on population growth, the number of megacities will continue to grow. A recent report by the Asian Development Bank predicted that by the year 2025 the number of megacities in Asia alone is likely to swell from nine to twenty, with several of them found in coral-rich areas.[40] Another study summed up the problem: "The majority of the coral reefs of the world occur in countries whose populations will double within the next thirty to fifty years."[41]

I am mulling over these grim statistics as Stefano and I return from our short visit to the corpse of Nyamuk Besar. My thoughts are as dark and murky as the bay itself. From the bow, we have an excellent view of the city. Thick plumes of greasy smoke roll skyward from all directions. It looks as if Jakarta is on fire.

"This must be what Beirut looked like during the civil war!" I shout to Stefano, and he laughs sadly and nods. As the boat speeds toward the sleek skyscrapers shrouded in a yellow haze, it strikes me that Jakarta really *is* burning, combusting slowly before our eyes.

11

"Either We Go Deep or We Starve"

There are only two threats to coral reefs: the needy and the greedy.

—Thomas Goreau, Global Coral Reef Alliance

The fundamental thing is that the greed created the need.

—Junie Kalaw, The Haribon Foundation

When the bomb goes off, we're still too far away to feel the blast. We don't hear it, either—although we are close enough for that—probably because the concussion is lost in the noise from Lida's twin forty-horsepower engines.

Lida Pet, a twenty-nine-year-old Dutch fisheries biologist, stands on the bow and waves as we approach the bombers. These aren't mere "Hi, how ya doin'?" hand flips: she makes large, loping movements with both arms, as if, out here on the Java Sea in the narrows separating Borneo from Sulawesi Island, she's run unexpectedly into long-lost friends. Her long brown ponytail blows in the wind like a flag proclaiming camaraderie. A smile is frozen on her face.

"I don't want them to think we're police," Lida whispers through her smile. As we approach, she signals her boatman to cut back the throttle, and we glide up to a large boat—the source of the bombs—with a dozen rough-looking men aboard. Four smaller outrigger canoes surround the main vessel, each with three or four occupants. The men are a mixture of Makassarese and Bugis, the two quintessential seagoing peoples of Indonesia, who had been plying these waters for hundreds of years before the seventeenth-century Dutch invasion. The Makassarese navy ruled the mighty kingdom of Gowa in the fourteenth century, with colonies spreading from what is now central Indonesia all the way to Singapore. The Bugis are descendants of pirates whose fearsome reputation gave rise to the word *Boogeyman.*

When our boat draws close enough, she calls to the men on board, still smiling, assuring them and reassuring them: we're friendly, friendly; no problems, no problems; hello, hello!

Lida has been based for more than a year in Ujungpandang, a city of nearly 2 million people on the southwest corner of Sulawesi, and she is fluent in Bahasa Indonesian. She is also beautiful—tall and lithe, blessed with luminous brown eyes and an actress's smile. Usually these two assets make her job of surveying the fishermen off Ujungpandang easy. But neither fluency nor beauty counts for anything in this particular situation.

The unsmiling men give us looks as sharp as knives.

"Don't take any pictures," she cautions me, without looking over. The warning is unnecessary, for I have no intention of even removing my camera from its bag. In fact, as we were approaching the boats, I nudged the bag itself under one of the seats, just in case its shape gave away the contents.

Lida's boatman is a serious-looking young man from Ujungpandang who is cultivating a thin mustache. She has him swing around and drift up beside one of the smaller outriggers. With a nod of her head, she indicates for him to offer cigarettes to the men in the boat. They take the little gifts and place them carefully in their shirt pockets. Each small boat has a compressor chugging away in the stern, sending air down a hose that disappears into the water. An older man with only a thin mantle of white hair on his head is friendly and chats easily with Lida, making small talk about the weather (it is the beginning of the *musim barat,* the season of the northwest monsoon) and, of course, fish.

What's going on here, under a canopy of gathering clouds, is called blast fishing. It is one of several techniques known as "destructive fishing practices." There's nothing fancy about blast fishing. You toss a small bomb into the water. After it explodes, you jump in and collect the fishes that have been killed or stunned. A few of the dead fishes float, and they are easy to skim from the surface. The problem is that four out of five sink to the bottom.[1] That's because the concussion from the blast usually ruptures the fishes' delicate swim bladder, a gas-filled organ that allows them to remain neutrally buoyant—to stay at a given depth without sinking or rising.

As Lida talks with the old man, a pair of hands suddenly emerge from the water and grasp the side of the small boat. The old man breaks off the conversation and grabs hold of the hands, hauling in a younger man, together with the bag of fish he has slung around his neck. The young man is slim, almost daintily built. His almost feminine physique makes the dark-blue dragon tattoos on his arms seem out of place. He pulls the breathing tube from his mouth and removes his goggles, rubbing his arms to warm up. His skin, which would normally be dusky, is tinged slightly purple, like the blush of a plum, from the cold. This is not surprising, since, according to the echo sounder mounted on Lida's boat, he has been collecting fish at a depth of just over a hundred feet. Lida estimates that he has been down between twenty and thirty minutes.

I ask her if she thinks he made a decompression stop on his way up. Lida smiles and shakes her head.

"That isn't done here."

Without making a short stop just below the surface after a dive of that depth and length, the young diver is courting disaster in the form of "the bends." Nitrogen, which builds up in the blood during a dive of this kind, is released during a slow ascent and especially during a decompression stop. Without these precautions, the bubbles of nitrogen can expand in the tissues, particularly in the joints, causing tremendous pain and, in severe cases, paralysis.

Not only do these men forgo a decompression stop, they actually race to the surface to dump their catch and, after moving to a different site, go back down, making several dives a day. Each time they descend, more nitrogen is added to their blood and tissues.

"Some know the dangers," Lida says when I press her on this, "but what can they do? If they don't hurry up with the fish, someone else will beat them to it. So they take the chance."

And some of them lose. Lida has seen men in fishing villages who are paralyzed as a result of this activity. Another researcher, a sociologist, tells me a chilling story about a village she visited in which an entire thatched hut on stilts contains nothing but young men, former blast fishers, who now lie paralyzed on woven mats, drooling and incontinent.

There are even more direct dangers from using bombs for "fishing." The most obvious is that if your timing is a bit off, the bomb will explode in your hand before you have a chance to heave it into the water. This frequently happens. A variety of "bombs" have been used, some safer than others, ranging from unexploded World War II ordnance (this was particularly common in Papua New Guinea[2]) to sticks of dynamite to homemade "bottle bombs" in which a glass beer or soda bottle is filled with gunpowder or a mixture of fertilizer and kerosene, and exploded using a homemade fuse.[3] This last type of bomb is probably responsible for the loss of more limbs than any other.

So many blast fishermen have lost hands and arms that a series of very dark jokes have sprung up about it. A group of foreign aid workers related a couple of the grimmer ones to me as we sat in the ostentatious lobby of Jakarta's Marco Polo Hotel:

What do you call a one-armed blast fisherman? Answer: experienced.

What do you call a blast fisherman with no arms? Answer: retired.

The effect of blast fishing on corals reefs is devastating. The explosion disintegrates living corals, destroying the reef framework and leaving craters up to sixteen feet in diameter.[4] I viewed this damage close up, earlier in the day when Lida and I dived a section of the reef where blasting was once frequent. The reefs are part of the Spermonde archipelago (Kepulauan Sankarang in Bahasa Indonesian), a necklace of some 160 coral islands, scores of patch and fringing reefs, and hundreds of acres of sea-grass beds and mangroves, strung along the southwest coast of Sulawesi in three fairly distinct rows. The islands lie in the Makassar Straits, a sixty-mile-wide body of extremely deep water separating Sulawesi from Borneo. The archipelago, which covers over six thousand square miles,[5] is the largest and most productive coral reef fishery in Indonesia.[6]

As we first entered the water, there was the usual Alice-through-the-looking-glass disorientation, for the Spermonde reefs are phenomenal, even by Indonesian standards, with what is probably the greatest coral diversity in the world. Some 262 species of hard corals have been identified in the Spermonde (approximately 75 percent of all species found in Indonesia), belonging to seventy-eight genera or subgenera.[7]

At first we swam slowly past intricate stands of branching corals, alternating with numerous colonies of tunicates, those small, primitive and delicate animals with eye-popping colors. Suddenly the colors were gone; an entire hillside of reef had collapsed into a pile of rubble. I could make out massive tabletops of coral that had been tipped over on their sides, half-buried in rubble, dead. I was reminded of photographs of the Murrah Building in Oklahoma City after it was destroyed by a terrorist's bomb: a lifeless tangle of pulverized gray stone.

As we continued swimming, we left this stricken zone and the colors returned. Living corals, and the profusion of life they shelter, once again dominated the reef. One study has estimated that it takes nearly forty years for a blast-damaged section of coral to recover from a single bomb.[8] But where blasting occurs continually—as usually happens—the outlook is far bleaker.

"Repeated blasting on the same reef can result in reefs which are little more than rubble fields punctuated by an occasional massive coral head," observes coral reef ecologist Mark Erdmann, "and coral recovery in these situations is unlikely given the unconsolidated, shifting nature of rubble."[9]

Blast fishing is far from new. It probably began soon after explosives themselves were first invented. Nor is the practice—which is illegal everywhere—limited to Indonesia. It has been blamed for the decline of fisheries throughout the world, from the Mediterranean[10] to Africa[11] to Southeast Asia. Blast fishing has been reported in forty countries,[12] and on at least 277 separate reefs.[13] The practice has been termed "rampant" in Tanzania, where dynamite stolen or sold illegally from limestone quarries and large gem mines ends up blasting reefs into rubble. The problem first arose in Tanzania in the 1960s, leading one scientist to state in 1968 that the reefs near the country's capital of Dar es Salaam would be virtually obliterated within a decade—a prediction that proved all too accurate.[14] Today, along that country's four hundred miles of reef coastline, "probably no reefs have not felt the impact of dynamite at one time or another—some reefs are totally destroyed (rubblized) by dynamite"[15] says Rodney Salm, coordinator of marine and coastal conservation activities for eastern Africa for the International Union of Conservation of Nature and Natural Resources.

In the Philippines, blast fishing has devastated an estimated two thousand square miles of what were once some of the world's most spectacular coral reefs.[16] On calm, clear days on one Filipino reef, researchers in the late 1980s counted as many as six blasts per hour.[17] In Australia, explosives used for fishing are referred to, with typically mordant Aussie humor, as "rapidly expanding bait."[18]

In Indonesia itself, blast fishing was already common enough by 1920 that the Dutch colonial government felt it necessary to pass a law against the practice.[19] Johannes Umbgrove, the Dutch geologist who wrote so extensively and so fondly about the corals of Jakarta Bay, noted in 1947 that "a large quantity of huge coral colonies . . . are blown up by native fishermen."[20]

Seemingly, wherever coral reefs exist in Indonesia, bombing has taken its toll. From the Mentawai islands, near Sumatra in the extreme west, to Bali and Flores in central Indonesia, to the coral reefs off the island of New Guinea in the far eastern reaches of the Indonesian Archipelago, blast fishing has been reported.[21] Even in the Bunaken Marine Park off Manado, in northeast Sulawesi, you can hear the dull thuds of homemade bombs throughout the day[22]—although the presence of divers has caused the bombers to move to more distant areas of the park.[23]

While individual subsistence fishermen do engage in blast fishing, the practice has become a big business, at least in the Spermonde. "It's enormously lucrative and enormously destructive,"[24] emphasizes Evan Edinger, a coral reef scientist from Ontario's McMaster University, who works on a joint project with Indonesia's Diponegoro University in Semarang.

Mark Erdmann, a friend and colleague of Lida's, spent three years in this area, researching destructive fishing practices. In a journal article, he described the kind of large operation Lida and I happened upon off Ujungpandang.

"Larger fishing vessels," he wrote, "embark on week-long voyages around the archipelago and into the Makassar Strait, and post comparatively large profits [$2,800–$4,650] for a full hold."[25]

That's a lot of money—particularly in a location where fishermen using traditional gear earn an average income of less than two dollars a day.[26]

That fact that the practice is illegal may stop a few people from blast fishing, but a more common result is that policemen have a new source of income: bribes to look the other way when bombs are tossed. Such payment—usually around a thousand to two thousand rupiah, or about a dollar—is called *uang rokok*, literally "smoking money." There is added incentive for policemen to accept a proffered bribe when they know that if they don't, the bomb that was created to kill fish is equally effective when used on men in uniforms.

We leave the bombers to their business, Lida pointing the general direction to her boatman, who needs little guidance. They have been making these tours of the coastal shelf off Ujungpandang four times a month for a year. Lida uses two different routes. The one we're on today is the shorter of the two, a small semicircle started at dawn, which can be completed in just a few hours. The longer trek extends over twenty-five miles and requires an overnight stay on an island out in the Makassar Strait.

She summarizes her work with the statement, "I'm trying to figure out fishing patterns through time and space," an explanation that fits a general rule I've discovered about scientific research: the shorter the summary, the more difficult and ambitious the project. This is evident as we continue on the route and stop at each fishing boat so that Lida can talk with the fishermen and record the size of each fish caught, the species (or, at least, the group), the total weight of the catch, the time spent amassing it, and the fishing gear used.

As we head toward the next boat, Lida explains that the bombers we saw were practicing a relatively benign form of blast fishing.

"At thirty-three meters they weren't bombing the reef, anyway," she says as she readies her equipment for the next fisherman—measuring board, scale, tape recorder. Unfortunately, the reason for the bombers' decision to "fish" in deeper water isn't any heightened environmental awareness or concern for the reef; it is that the combination of blast fishing and increased fishing pressures from the rising human population has decimated the reef-fish population. Even twenty years ago, say local villagers, they could start a pot of rice cooking, go out and catch enough fish for the meal, using traditional gear such as hook and line, and return before the rice was done.[27] Now it would take several hours to catch that same amount of fish.

Today there simply aren't enough reef fish left to make blast fishing economically viable. This is the pattern—heartbreakingly typical here and elsewhere. Blast away until the reef fishery is effectively wiped out, and then move on to open-water fishing or to other, more pristine reefs, and destroy them. It has been repeated throughout Indonesia, in the Philippines, and in many other countries.

For a couple of hours we cruise between fishing boats, most of them small outriggers holding single men, sometimes a father with his young son along. They use small nets or simple hook-and-line gear, and chat easily with her boatman as Lida works. They laugh and grin courteously when I attempt conversation with my minute vocabulary of Bahasa Indonesian. One would have to work very hard to dislike the people of Ujungpandang and the surrounding islands. There is a general and pervasive friendliness that feels genuine wherever you go.

The clouds continually build and then break apart. When they do separate for a few minutes, sunlight pours through, a light so intense that it fills the air around us like the clanging of a large bell. Against this sun, and the anticipated rain, many of the men wear the conical hats of woven palm fronds typical of Southeast Asia. The low-built city of Ujungpandang is always in the background, faintly visible through the blue urban haze, its many minarets rising up here and there from neighborhood mosques. Behind it all, dull-green mountains loom over everything.

We pull up beside a small outrigger, where a man cooks a fish on a small wood grill. On the side of his boat, the name *Hunter-Hunter* is painted in bright blue letters. The man wears a Pittsburgh Steelers sweatshirt, and after Lida's boatman hands him a cigarette he smiles and admits that he thought at first we were police. He isn't doing anything illegal, but even so, if "smoking money" isn't forthcoming, fishermen are often threatened with arrest for some infraction or other. Corruption (*korupsi* in Bahasi Indonesian) is a part of life throughout Indonesia, as it is wherever authoritarian governments rule. Occasionally, Suharto's government will initiate another in a long-running series of "anticorruption" campaigns. These crusades have a comic-opera quality about them: Ministers feign surprise, as convincing as Inspector Louis's shock at finding gambling in Rick's Café Américain. Headlines in the heavily censored press decry "the reprehensible actions of a few parasites." A few small fry are tossed into jail or reprimanded and lose their pensions. When the dust settles, life, and graft, returns to normal. And why shouldn't it, with Indonesia's First Family also serving as the uncontested First Family of *korupsi?*

Lida asks the fisherman for his catch, and he happily hands over a small plastic bucket. She dumps them into a bamboo basket, where they still glitter, the iridescent yellow lines on their sides slowly turning dull as life leaves them.

She measures and records their length and grouping, speaking in Dutch into a small tape recorder. As she works, her boatman shares a smoke with the fisherman, who seems amused by the process. At least it breaks up the monotony of his morning.

When she's done, Lida pours the fish back into the plastic bucket and hands it back, politely inquiring how fishing is today, although she can already tell from her inventory.

Her translation of his answer is, "It sucks."

"Why go out, then? Why not just stay home in bed?" she asks.

The man shrugs. "There's always a chance of catching something," he answers. "Besides, I had a fight with my wife last night."

Lida laughs, thanks him, and we speed off to the next boat.

This one is different from the others. It is larger, and supports a complex bamboo structure on either side, which hold the *bagan*, or lift net. The boat and its crew of five goes to sea for three weeks at a time, returning each full moon to their island, Barang Lompo, which is midway in the archipelago, in the second strand of islands. The men fish at night, using lights to attract their prey (which explains why they take a break around the full moon, when bright silver light shines evenly across the ocean's surface). During the day they sleep, repair nets, smoke Kretek cigarettes, talk, and sell their catch to middlemen who pick up the fish in smaller boats and sell them at the marketplace in Ujungpandang. It is midday. With typical Indonesian hospitality they invite us aboard to share their lunch of rice and Indian mackerel.

Later, feeling pleasantly groggy in the hot sun after eating, we head off to complete Lida's circuit. Off a coral island dotted with huts and coconut trees, we approach a low, simple structure, a collection of nets fastened onto planks of wood, floating on blue plastic barrels. A flimsy shack of woven palm fronds completes the arrangement. Three magnificent fish swim placidly about in the net pens. Each fish is large—a foot and a half to two feet long—and exquisitely colored. Two are of a species called saddleback coral grouper (*Plectropomus laevis*), also known as coral trout; several black "saddles" run down their snow-white sides as if someone has spilled ink on them, and they have bright yellow tails and fins. But the third fish is the most striking. It is probably a coral trout, too, and may even be the same species as its pen mates, but it's hard to tell, because this one is an albino. Its black, shiny eyes rest like marbles on a field of white. The fish is so solidly built and so dazzlingly white that it is hard to look at it and not think of a miniature Moby Dick.

This holding pen doesn't have the drama of the blast-fishing site, and yet the ramshackle structure gently bobbing on the waves represents an even greater threat to coral reefs and their fisheries than that posed by the bombers. For while blast fishing still continues, there is evidence that it will fade in significance before long, the bombs having destroyed the large aggregations of fish needed to make it worthwhile. These nets are the picture of devastation yet to come.

Nearly all the fish we've seen today are destined for the crowded and noisy markets of Ujungpandang. But not these three in the holding pen. Instead, a large vessel, usually with Chinese characters on the side, will come calling for these exquisite fish in a week or two. With their twenty-ton holds filled with live reef fish, these ships will head north, sailing for Hong Kong, Taiwan, or Singapore. There, the still-living fish will be unloaded and taken to gourmet Chinese restaurants where they will be transferred, one last time, into large aquaria. At night, wealthy diners will examine these exotic and brightly colored fish, remarking to each other on their size, beauty and vigor. When they see one that strikes their fancy, they will point to it and the waiter will net the fish and take it into the kitchen, where it will be quickly steamed and delivered elegantly and with great fanfare to the diner's table.

It has been called the latest fad among the new wealthy elite of these countries: dining on recently living coral reef fishes. But to term it a fad is to understate its significance. Call it, instead, a cultural ritual, for the ability to purchase beautiful, rare, and enormously expensive reef fish announces to the world that you are someone to be reckoned with. To gather with friends and eat the delicate flesh and brightly colored skin of reef fish sends a message: you are a man of status. Dining on the lips of a Napoleon wrasse—a delicacy that can fetch $225 a plate—is the height of Asian (particularly Cantonese) conspicuous consumption.

The arrival of an exceptional fish is cause for celebration in Hong Kong, where between 100 and 150 companies compete for the live reef fish trade.[28] In November 1996, the *South China Morning Post* reported on the fuss made over a rare giant grouper, nearly ten feet long and weighing over five hundred pounds.[29] A restaurant paid over ten thousand dollars for the fish, which was probably fifty years old. "We have asked an expert to choose an auspicious day for chopping up the grouper," said the proud restaurant manager. "And we'll hold a small ceremony before our chef wields his chopper."[30]

Lida Pet is a reluctant expert in this field. It is, in fact, one reason I sought her out in the first place, after reading a journal article she coauthored on the live reef fish trade.

After our trip around the Spermonde Archipelago, we climb into Lida's four-by-four and, with Joan Osborne blaring from her car stereo, head for her house near central Ujungpandang, where we can discuss this new and growing threat to coral reefs. After weaving through the tangle of *becak*s, we are, at last, seated in her spacious living room, ceiling fans slowly tracing circles above us, a welcome refuge from the noise, heat, and dust of the city. The room is filled with expatriate trappings: a large bronze Buddha's head from Sri Lanka rests on the coffee table. Finely woven rice baskets hang on the wall. But scattered throughout the room are reminders that Lida is not your typical expat. If you allow your eye to rove, you keep encountering gallon jars around the room. Each one contains a baby shark preserved in alcohol. On a bookshelf is a row of jars containing a black-tipped shark, a nurse shark, and a shark-ray. "It's not a true shark," Lida hastens to explain, "but, well, I just like it." Across the room, curled around in its jar like a cat settled in for the night, is a beautifully mottled bamboo shark.

As we sip on Coca-Colas (Lida admits, shamefaced, that she is addicted to the stuff), I ask her to tell me about the live fish trade.

"Yes, well, the practice is terrible," she says. "And they have to go farther out now to get the fishes, for three weeks at a time."

To obtain these fishes, Lida explains, divers use discarded plastic squirt bottles filled with a solution of sodium cyanide. They dive on the reef, searching for one of the "target species," usually a rock cod, a large grouper, or a coral trout like the ones we saw in the holding pen. When they locate their prey, they stun the fish with a squirt of the poison, or chase it into a coral hole and squirt the cyanide solution into it. In the latter method, the coral is then pried apart with a metal bar to get at the stunned fish. The damage caused by this is obvious, but even when a pry bar isn't used, the cyanide in the water kills corals. And other fishes. And nudibranchs, tunicates, sponges, anemones, crabs, shrimps, isopods, jellyfish, snails, octopuses, sea turtles, and all the other millions of creatures that make their homes on the reef.

"Of course, that's the tragedy of this," Lida says, the disgust evident in her voice. "The poison kills everything. *Everything.* It's stupid."

Just as in blast fishing, the independent small fishermen using cyanide are only part of the problem. In addition, Lida and her coauthor wrote in their journal article, there are also "well-organized teams of divers working from large 'catcher' ships equipped with 6–10 fiberglass dinghies and live-hold tanks that can accommodate 1–2 [tons] of live fish."[31]

The money paid to local fishermen for their catch of live fishes is paltry beside what customers in Hong Kong fork over for "the ultimate dining experience." Still, they receive significantly more for live fishes than they do for dead ones of the same variety—up to twenty-five times more.

"They can earn up to five hundred dollars a month by selling live reef fishes," Lida says. "That's three times what lecturers in a university here make."

Driven by these large sums, the size of the live-reef-fish trade is enormous, but accurate figures are impossible to obtain, due, in part, to the fact that cyanide fishing is widely illegal. The most comprehensive study of the trade to date, produced for The Nature Conservancy, estimated the total annual Asian trade in wild-caught live reef fishes at between eleven thousand and sixteen thousands tons,[32] with Indonesia providing over half of the total[33] and earning the country $200 million a year.[34] Ujungpandang, the shipping capital of eastern Indonesia, has quickly established itself as the central transport point for the trade.

The industry has become so lucrative that air shipments of fishes are now replacing the cheaper but slower boats, which take a week or more to make the journey to Hong Kong. Ujungpandang's recently expanded international airport will likely increase the trend toward air-freighting live fishes—at least for a while. As Lida pointed out, cyanide fishers are already having to travel farther in search of their precious quarry, having virtually wiped out local adult populations of the most sought-after species. The process, which has been rightly compared to strip mining,[35] removes the largest, most fecund females, depleting breeding stocks and nearly ensuring the demise of entire species throughout large areas.

Exports of certain favored reef species such as the humphead wrasse have already declined,[36] and industry representatives themselves estimate that by the turn of the century, the live reef-fish trade will no longer be financially viable in Indonesia, owing to decline of the fisheries.[37]

The premium attached to "exotic" fishes leads to what amounts to an ecological death spiral, in which the rarer a species is, the more valuable it becomes, and so more sought after. "Being endangered actually seems to spur demand," observed one Hong Kong resident.[38]

It is hard to imagine a more effective way of driving a species into extinction. And that is precisely what is happening.

"If we don't act soon," a reef biologist warned recently, "global extinctions may follow the many local extinctions that have already occurred."[39]

Again, as in blast fishing, cyanide fishing leaves a human health crisis in its wake, with fishermen themselves becoming victims of the methods they practice. As the shallow reef areas are "fished out," fishermen are traveling not only farther but deeper to find target species. They often employ the type of primitive breathing equipment used by blast fishermen (sometimes utilizing old paint sprayers for air compressors), with little or no training on decompression techniques, and suffer the same consequences. In 1993, in a single fishing village in the Philippines, there were thirty serious cases of the bends among two hundred divers, ten of those cases resulting in death.[40] When a government extension worker warned the villagers of the risks involved, he was told, "Either we go deep or we starve."[41]

Nenny Babo, founder and director of the Sulawesi Natural Resources Conservation Information Center, a private environmental group, tells me much the same thing over dinner at Lida's: "You talk to people who use blasting or cyanide and they say they have to do it to eat."

If you discuss what's happening to the Spermonde reefs with scientists like Lida Pet or activists like Nenny Babo, at some point the conversation will inevitably turn to poverty. In fact, economic issues are a constantly recurring theme in any serious discussion about threats to coral reefs in Southeast Asia. Although things are changing, some environmentalists and scientists in the developed world still talk about the destruction of coral reefs as if it were happening in an economic and political vacuum. In the developing world, however, there is a general understanding (born of necessity) that economic, social, political, and environmental issues form a single, inseparable braid. The entire complex of issues, unwieldy as it is, must be considered as a whole if reefs are to survive.

"As long as there are poverty-stricken people that are sustained by the sea, and as long as there is demand by the rich and wealthy for 'luxury' fish, dynamite and cyanide will continue to send our coral reef to irreversible degradation," observes Michael Aw, a veteran underwater photographer of coral reefs. "What can these governments offer the poor fishermen who are making considerably higher wages (at the risk of their lives and health) using cyanide to catch a few species in high demand?"[42]

That perspective is shared by Dr. Steve Oakley, a professor of environmental science at the University of Malaysia's Institute of Biodiversity and Environmental Conservation. "Outsiders supporting reefs, education, enforcement are all part of the solution," he says, but warns that "the biggest problem is poverty combined with no ownership of the resource."[43]

These strands of poverty and political marginalization are cited as primary factors leading to the continuing decline of sea turtles—even after international trade in marine turtles has been, if not eliminated, then at least drastically curtailed.

"Demand for international commerce is now an insignificant factor," conclude the authors of *The Biology of Sea Turtles*, "but has been replaced by increasing demand for subsistence and local markets by indigenous people, whose population increase has often not been matched by an increase in *real wealth* or *political opportunity*. [My emphasis.]"[44]

This far more complex way of thinking about "environmental" problems (that they aren't *really* environmental at all, in the sense that they are divisible from other issues) represents an important change—potentially the most profound shift in our thinking since Rachel Carson first sounded the alarm about environmental degradation more than three decades ago.

There are two components to this new understanding. First, as Sylvia Earle, marine biologist and former chief scientist for the National Oceanic and Atmospheric Agency, has pointed out, humans are not separate from nature, but part of it. We are, to continue the metaphor begun with coral polyps, a single strand in the braid of life. And, second, given that fact, the full range of human activity—including the role of government, economic, social, and religious structures—must be considered when attempting to reduce anthropogenic threats to the coral reef. To come to terms with nature, we must come to terms with ourselves—in all our multifaceted, convoluted, messy ways.

Of course, neither of these ideas is new. The first one has been repeated in so many different ways, by so many writers, ancient and modern, that it is the Methuselah of environmental clichés: "I am one with nature." For all the repetition, however (and perhaps because of it), this putative understanding of our place within the natural world remains merely a platitude, cotton candy for the soul—a commonplace of callow poets and, in its most cynical manifestation, of television advertisements for oil and chemical companies. If ideas based on this philosophy were raised in policy-making circles, they would be greeted with embarrassed silence—as if our intimate relationship to our environment were irrelevant to the actions of individuals, governments, corporations, and other institutions. Signs of this dualism are everywhere. My personal favorite (in the sense of loving-to-hate-it) is the gas-guzzling "sport utility vehicle" used in cities and suburbs to drive errands easily accomplished on foot or by bike or mass transit—and with a bumper sticker beseeching us to "Love the Planet" slapped on. Even marine ecologists who should know better unwittingly fall into this dualistic trap, labeling threats to corals reefs as either "human" or "natural"—as if humans really were, somehow, intrinsically outside of the natural order.

The task, then, seems simple enough: to start acting according to our deepest and best beliefs. But of course nothing could be harder for humans. If it were easy, the entire self-help movement, to say nothing of religions, would go out of business overnight. And nowhere is this task more critical, or more difficult, than on coral reefs and in their associated ecosystems of sea-grass beds and mangrove forests. Critical, because reef fisheries are the lifeblood of hundreds of millions of people in areas of burgeoning populations. And difficult because those same areas are some of the poorest in the world, many lacking the basic infrastructure (such as sewage treatment) that buffers the environment from human activity. In addition, the traditional social order in those areas has been abraded by a variety of factors ranging from authoritarian governments to the global market economy.

For all of these reasons, and more, coral reefs are the proving ground for mankind's ability to "come to terms with nature"—including our own.

Coral reefs aren't just beautiful, they're also one of the most complex ecosystems on earth. Their biodiversity is rivaled only by tropical rain forests. At least 10 percent of coral reefs worldwide have already been destroyed, and another 30 percent may soon meet the same fate. (Michael Aw/Ocean N Environment)

What look like grains of sand inside this polyp are actually single-celled algae that live in a symbiotic relationship with coral. Many coral species depend on the photosynthesizing algae called zooxanthellae to provide up to 90 percent of their nutrients. (Clay Cook/Harbor Branch Oceanographic Institution)

Like most corals, the beautiful *Tubastraea faulkneri* (left) keeps its polyps withdrawn during daylight hours. The huge, brilliantly colored polyps emerge at dusk (right) to feed on zooplankton. (Michael Aw/Ocean N Environment)

Pink sperm-and-egg bundles are released by a colony of the coral *Montastraea franksi* eight days after the August full moon. Billions of individual coral polyps from millions of coral colonies synchronize their annual nighttime spawning, creating huge gamete slicks in the oceans. (Greg Bunch/gb undersea)

The stinging tentacles of a sea anemone *(upper right)* protect these brilliantly colored clownfishes from predators. The fish larvae are coated with the anemone's chemical signature so that the anemone is unable to distinguish between the clownfish and its own tentacles. This way, the fish doesn't trigger its host's stinging cells. (Michael Aw/Ocean N Environment)

A green moray eel *(center right)* is being "cleaned" by the small, striped goby fish near its eye. Gobies and other small fishes set up cleaning stations at specific locations—for example, on top of a large coral head. The larger fish visits the station and performs certain movements or changes colors to signal its desire to be cleaned. (Allen K. Wicks/WET Adventures)

A variety of tunicates, or sea squirts *(bottom)*. The large, bluish, vase-shaped tunicates are individual organisms; the smaller, red, strawberry-shaped ones are colonies of several tunicates. These filter feeders are our distant relatives. Their primitive nerve cord (a simplified version of our spinal cord) exists only for a few hours during the larval state. (Michael Aw/Ocean N Environment)

Crinoids, or feather stars, are the most ancient members of the phylum Echinodermata, a large group that includes sea stars (sometimes called starfish). Rachel Carson called echinoderms "the most truly marine" of all invertebrates, because not one of their approximately 5,000 species has made the transition from sea to land. (Michael Aw/Ocean N Environment)

Reef-building "hard" corals assume myriad forms near the ocean's surface, where life-giving sunlight can reach the algae living within the corals' tissues. This part of the reef is continually growing upward, matching rising sea levels as corals secrete layer after layer of calcium carbonate. (Michael Aw/Ocean N Environment)

Farther down the reef face, where only dim sunlight penetrates, hard corals give way to beautiful and delicate non-reef-building "soft" corals. (Michael Aw/Ocean N Environment)

This threadfin butterfly fish exhibits several characteristics common to reef fishes. Its "skinny" profile allows it to hide in the reef's many crevices and holes, while its brilliant colors make the fish instantly identifiable to others of the same species. (Michael Aw/Ocean N Environment)

This "mimic octopus" was recently discovered off the island of Sulawesi in north-central Indonesia. It is named for its ability to assume the color and shape of a large number of marine creatures, including crabs, sea horses, flatfish, lionfish, and shrimp. (Susan Ritman)

Exhausted after a night spent laying and burying a clutch of some one hundred eggs, this massive green sea turtle (her shell is over a yard long) rests for a moment at dawn before returning to the lagoon surrounding Heron Island on Australia's Great Barrier Reef. (Osha Gray Davidson)

Mangrove forests prevent land-based sediments from drifting out to coral reefs, and protect coastal residents from hurricanes and typhoons. (Joseph W. Dougherty/WET Adventures)

Mangrove forests throughout the tropics are being cut down at an alarming rate, primarily to build shrimp farms for the Japanese, American, and European markets. Fully half of all mangrove forests in Southeast Asia, like this one in Thailand, have been destroyed over the past fifty years. (Alfredo Quarto/Mangrove Action Project)

Poor agricultural practices and logging in Costa Rica have caused sediments to flow into rivers and then out to sea, where they cloud coastal waters, killing corals. Sedimentation is one of the leading threats to corals around the world. (Joseph W. Dougherty/WET Adventures)

Algae-eating fishes have been called the reef's immune system, since corals and algae often compete for the same space. Overfishing often allows the algae population to explode, smothering corals. (Michael Aw/Ocean N Environment)

Blast fishing, using dangerous homemade bombs, is often the last resort of subsistence fishermen struggling to feed their families in overfished waters. Before being dynamited, this was a healthy reef off the island of Borneo. (Michael Aw/Ocean N Environment)

A white limestone skeleton is visible through the transparent coral tissues on the right. After the zooxanthellae algae have abandoned the coral, the result is a frequently fatal condition known as bleaching. Some scientists believe that many mass coral bleaching episodes in the last decade may be related to global warming. (Michael Aw/Ocean N Environment)

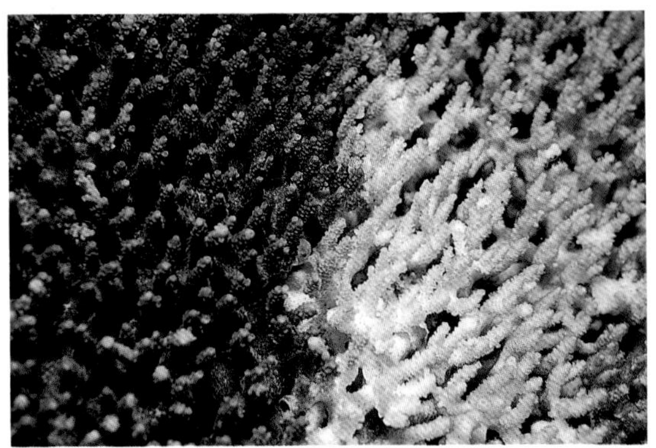

A Hawaiian green sea turtle with multiple fibropapilloma tumors. The tumors are benign, but infected turtles usually starve to death once the growths cover their eyes and mouth. Water pollution may suppress the turtle's immune system, leaving it vulnerable to this disease, which has reached epidemic proportions in Hawaii, the Florida Keys, and other sites. (Ursula Keuper-Bennett)

The burgeoning "live reef fish trade" is pushing several species of coral reef fishes toward extinction. Wealthy Asian diners pay up to $225 to dine on the lips of a single Napoleon wrasse. This species is shown here in a holding net on a central Indonesian reef, awaiting live shipment to restaurants in China. (Mark Erdmann)

If reefs are to survive in many developing nations, the pressures on fishing must be relieved. In the waters near Bali, Indonesia, former fishermen have found a sustainable alternative livelihood by "farming" seaweed that is then sold to Japanese cosmetic and pharmaceutical firms. (Osha Gray Davidson)

12

The Apo Scenario

> I am terrified for the future of my children. How can they survive in this kind of situation? What can they look forward to? . . . But in the end, we must keep on hoping and working.
>
> —Leonor Briones, Freedom from Debt Coalition

If humanity is to make the quantum leap from aphorism to action, it will be in large part thanks to people like John McManus.

McManus works for the International Center for Living Aquatic Resources Management (ICLARM), a research organization based in Manila that strives to improve the productivity of fisheries and other marine resources in developing countries. I had been reading scholarly papers by McManus, and hearing others speak about him in reverential tones for several months, before I had the chance to sit down and talk with him in his office in Manila. The word his peers use most frequently when asked to describe McManus is "visionary." He is known for performing the scientific equivalent of the old magician's trick, pulling a rabbit out of a top hat. The hats McManus reaches into are highly complex concatenations of data sets, and instead of rabbits he pulls out—ta-da—elegant, original, and comprehensive views of what is happening in the real world. Perhaps his greatest magic trick of all, McManus can explain to laypeople *why* what's happening is important.

As the elevator doors slide open, McManus is already there to greet me, his hand outthrust, an endearingly goofy smile on his face. He leads me to his comfortably crowded office on the third floor of an office building in Makati City, an upscale financial and shopping district of metropolitan Manila. It is the end of November and the extreme humidity here, only 15° north of the equator, makes walking seem closer to swimming. I return to my hotel three times each day, just to change my sweat-soaked shirt. Improbable as it seems to someone from the frigid northern hemisphere, Christmas decorations are everywhere in this muggy cityscape. The tinsel and the nativity scenes, I learn, have been on display since early October. Windows contain star-shaped lanterns called *farols*. Loudspeakers ooze syrupy renditions of—improbably enough—"I'm Dreaming of a White Christmas." The seriousness with which Filipinos take Christmas shouldn't be too surprising. This is a country, after all, in which men celebrate Easter with bloody displays of self-flagellation and, in some areas, by having themselves nailed to a cross.

All the festive gear wishing Peace on Earth, Good Will Toward Men, seems a bit incongruous with the army of security guards assembled in front of nearly every building in Makati, however. Each man bears a small arsenal that would make a member of the Michigan Militia drool with envy. In addition to sidearms of every description (some guards wear a holster on each hip, just as in the old American West), most carry pump-action shotguns, with extra shells lining their belts. As I sign the visitors' book in the poorly lit entryway to McManus's building, three young men stand around in the gloom watching me, shotguns cradled in their arms, handguns at the ready.

Once inside his office, McManus puts me immediately at ease. During our long talk, he laughs easily and frequently, and often loud enough to rattle the mobile of brightly colored reef fishes that hangs from the ceiling just behind him.

For all his gregariousness, however, McManus's message is sobering.

"Seventy percent of the world's fisheries are *already* at or beyond carrying capacity," he points out. "Entire sea-grass beds, nurseries for many reef species, have been wiped out. If you want to use the term *ecological catastrophe* somewhere, it applies here."

McManus can speak for hours, giving dozens of examples of destructive fishing practices that have resulted in the current grave situation for reefs and reef fisheries.

"Cyanide fishing has a long history of use in the Philippines," he says, "predating the recent growth in the live-reef-fish food trade." Fishermen have used cyanide to supply ornamental fish for the international aquarium trade, causing the same type of destruction as associated with the live food trade, but on a smaller scale. Far worse, he has witnessed men emptying fifty-five-gallon drums of cyanide directly onto reefs, a "fishing" practice that wipes out every living creature in the area, including corals.

"Drive-in techniques," in which groups of people scare fish into large nets, has also taken a toll on fish, corals, and humans. The most widely practiced form is called *muro-ami*. McManus explains that *muro-ami* has been practiced on a small scale for some time, often involving many people, particularly young boys, from small fishing villages. After a net has been set up in a large semicircle, the boys swim toward the open end of the net while dragging "scare lines," long ropes with rags tied to them and weighted at the bottom. Filled with thrashing fishes, the net is hauled up. Even at this modest level, *muro-ami* has destroyed large sections of coral reefs throughout the Philippines, because the weights used to keep the "scare lines" vertical end up breaking corals. But in the 1970s, *muro-ami* went commercial, involving many large vessels with hundreds of divers (usually underpaid young boys) driving enormous numbers of fishes into huge nets. Not surprisingly, the problems have increased with the scale of the operation. The increased efficiency and sheer size of the large commercial operations have led to severe overfishing. The hundreds of weighted scare lines have shattered many coral reefs into rubble (reducing the possibility for future replenishment of coral reef fishes). And the use of child labor, though on the decline following a government crackdown, remains a problem.

One of McManus's current concerns is trawling, an activity not usually associated with reef destruction because trawling boats, dragging their huge nets behind them, traditionally stay in deeper water.

"We need to rethink this problem," says McManus, in part because reef research has traditionally focused on shallow-water reefs. But deep-water coral communities are rich ecosystems themselves, and are increasingly exploited by trawlers.

"There is probably an order of magnitude more coral in uncharted areas than in charted ones," he says. "Sure, they're slow-growing at that depth, but that doesn't mean they're small or unimportant. Some people will tell you that the bottom is a silty, sandy place. Desertlike and flat. Well, *now* it is, after trawlers have leveled everything."

But before trawling became so widespread, these areas were lush, thriving ecosystems, "covered with sponges the size of this desk"—a good six feet by three feet or more—"in patches the size of this building"—half a block long. "And every sponge is home to hundreds of other organisms," McManus reminds me.

In the past, trawlers avoided coral areas because the benthic (seabed) structures there could shred the giant nets they used. But those areas are currently being made "trawlable." Vessels criss-cross them, dragging chains and other heavy objects, destroying corals and other bottom dwellers.

"Not only do trawlers do devastating damage to the coral areas themselves, and deplete fishing stocks," says McManus. "They push small-scale fishers into destructive fishing practices."

A key element to understanding all of these problems, McManus explains, is found in the concept of Malthusian overfishing, a principle originally defined as "occur[ring] when poor fishermen, faced with declining catches and lacking any other alternative, initiate wholesale resource destruction in their effort to maintain their incomes."[1] Examples include blast and cyanide fishing.

"The problem isn't just population growth," McManus emphasizes, "but that's an essential part of it. And probably the easiest to see. After all, most coral reefs are in countries where the population will double within thirty to fifty years. But the other part of the equation, what's often forgotten, is the increasing levels of *inequity* in these places."

This is familiar terrain to McManus, who has written frequently and passionately over the past decade, trying to get scientists and policy makers alike to wake up to the complex realities involved in tackling the life-and-death issue of Malthusian overfishing.

Fisheries management, he has written, "is primarily an applied social science":

> Most of the emphasis in literature and application in fisheries has been centered on the natural sciences, and far less has involved social sciences such as economics, sociology, social psychology, anthropology, or political science. This imbalance in approach has been especially problematic in coral reef fisheries, wherein the majority of attempts to effectively implement management plans have failed because of noncompliance by part or all of the target human populations.[2]

And, McManus adds, when the "target human populations"—i.e., fishermen—don't comply with experts' recommendations, the experts have a predictable explanation: "A fisheries person says that people were 'silly' for not taking the advice. But you have to remember that when we talk about management, we're talking about *human* behavior."

Which means that scientists and managers must break out of their narrow boxes and grasp the basic principles of other disciplines—many outside of the "hard" sciences. Again, this process is further along in the developing world than in the so-called "advanced" world.

McManus illustrates this point by relating a conversation he had at a recent international symposium.

"During a session," he recalls, "a high-ranking person at a major institution in the 'developed world' leaned over to me and asked who he could talk to about community structure. I looked around the room and pointed to Dr. X. And this man said, 'But isn't he an ichthyologist?' Well, yes, he was. But he was *also* an expert in the structure of the fishing community in which he worked. In fact, he *had* to be, to do his work."

It may be coincidental that the marine ecosystem with the greatest biodiversity has spawned this call for "scientific diversity." But I suspect there is a more pragmatic reason at work. Ever since the young geologist Charles Darwin first investigated living coral reefs, the field has been the most interdisciplinary of all scientific pursuits. Geologists commonly mingle with ichthyologists. Zoologists trade notes with botanists. Molecular geneticists coauthor papers with ecologists. The braided nature of corals, and of the reef/mangrove/sea-grass triad itself, has required greater cooperation between scientific communities which generally keep pretty much to themselves. It's more than likely that the new emphasis on combining "hard" and "soft" sciences builds on this rich historical groundwork.

McManus isn't the only one preaching the gospel of interdisciplinarianism. At a major conference on coral reefs, held in the Philippines in 1995, a French scientist named Bernard Salvat told the gathering:

> [T]he science of coral reefs must include not only the natural sciences such as biology and geology, but it must also include social science to develop sociological and economical research related to reefs and reef uses. Links between coral reef environments and some human societies have to be considered within the context of the traditions and cultures of these populations. We must also consider the acculturation of these societies which is occurring with the westernization of these countries.[3]

And at the Eighth International Coral Reef Symposium in Panama, Australian Chris Crossland spoke of the need to "collapse the triangle"—comprising marine park managers, users, and scientists. "We can't keep talking in jargon all the time," Crossland insisted. "Marine science is irrelevant if you don't put it to use."[4]

And yet the traditional walls remain largely in place, separating scientific disciplines from each other, separating the "hard" from the "soft" sciences, marine park managers from scientists, and scientists from the public. The plain fact is that scientists *do* speak in jargon most of the time, and hardly a day goes by on which a pundit doesn't complain about the very real and widespread problem of "science illiteracy" among the public.

Even John McManus, one of the more optimistic individuals you could hope to meet, doesn't minimize the difficulties of the task—if coral reefs are to be saved.

"If you stop doing all the negative things," he says emphatically, "reefs *will* come back. But the problem is that people *haven't* stopped doing the bad things. Here's the principle: You have to turn off the faucet *before* you start mopping the floor. And turning off the faucet implies massive changes in politics and economics. Coral reef people alone can't do this."

He sighs and grows pensive.

"It's very difficult to imagine these changes happening," he says, continuing down this rather bleak road. "Fisheries are being depleted. Land is running out." He shrugs and sighs again.

Outside, the tropical sun is descending, leaving behind a humid darkness that seems to breathe on its own. All over sprawling Manila, lights are coming on. There are the humming neon lights of Ermita, ground zero of the city's flourishing disco scene, where wealthy businessmen and women dance and drink the night away. A bit farther north, across the Pasig River and beyond Chinatown, bare light-bulbs, often connected illegally to power lines running overhead, illuminate the vast, congested slums of Tondo. Even farther up the coast, in the many fishing villages along the South China Sea, children sit below sputtering kerosene lanterns, packing chemicals into glass bottles and inserting fuses, preparing the bombs their fathers will use for tomorrow's fishing.

Lights are going on to the south as well, throughout the seven thousand islands of the Philippine Archipelago. Some five hundred air miles southeast of Manila, a lighthouse on tiny Apo Island clicks on, illuminating a beacon that is seen by boats far out in the Mindanao Sea. For years the fishermen of Apo had practiced the same destructive techniques as on other islands: *muro-ami* and cyanide and blast fishing. And the result was the same: destruction of the surrounding reefs and ever-smaller catches of reef fishes. Even to bring in these dwindling amounts of fishes, fishermen were finding they had to travel farther all the time.

But today things are different on Apo. Something has happened that provides a small measure of hope that, to use McManus's analogy, the faucet can indeed be turned off and the floor mopped up. Call it the Apo Scenario—a hopeful counterbalance to the devastation of Jakarta Bay.

The changes began in 1979, when a group of scientists from Silliman University, on the nearby island of Negros, started visiting Apo. The villagers told the scientists about the problems they were having catching fish, and the scientists talked about the concept of "sustainable use." By 1982 the people of Apo, working with the scientists from Silliman University, set up a marine reserve on a small portion of their 262-acre reef.[5] Within this twenty-acre zone (just 8 percent of the total), fishing of any kind was off limits, a protected status that was enforced by islanders themselves.

The results were discouraging at first. Between 1982 and 1985, scientists monitoring fish abundance found no evidence of increase—inside or outside of the reserve.[6] Part of the problem may have been lax enforcement of regulations. Even in 1984, at least some fishermen from Apo fished within the reserve boundaries at times.[7] Regardless of the reasons, villagers began to wonder if "sustainable use" was really the answer to their problems. Still, because the scientists from Silliman University had built a good relationship with the villagers on Apo, the plan continued. Then, between 1985 and 1986, the number of species within the reserve slowly rose, as did the abundance of certain types of fishes.[8] By 1988 the number of large predatory fishes within the reserve had increased over three and a half times.[9] By 1993 the density of those fishes had increased nearly eightfold.[10]

So far, so good. By excluding fishing in a particular area, stocks had grown. That was a significant victory for Apo islanders, and for proponents of fishing reserves. But it left an even more important question unanswered: Did the increase within the reserve lead to an increase in fishes *outside* the reserve area, in neighboring waters—what fisheries managers call a "spillover effect"? After all, even exponential growth in the fishing stock isn't sustainable in the long run if local fishermen aren't allowed to benefit from that increase. For a small marine reserve like Apo, where enforcement is entirely in local hands, the cooperation and support of fishermen is critical.[11]

Again, time was the crucial factor. Remember that it took two to three years for scientists to detect any increase in fishes within the sanctuary. It took another several years to see any "export" of fish biomass beyond the borders of the reserve. Finally, between 1991 and 1993, after nine to thirteen years of the reserve program, scientists found increased numbers of fishes outside the reserve, with the largest increases found in waters closest to the reserve.[12] Though some scientists discount the opinions of fishermen about fluctuations in catch sizes (calling them "anecdotal"), the people who depend on the fisheries for their livelihood are probably best equipped to assess the effectiveness of reserves. In 1986, eleven of twelve Apo fishermen surveyed said that their catch (taken from outside the sanctuary) had increased since the reserve was established. By 1992, all twenty-one fishermen surveyed reported that their catch had doubled since 1985.[13]

Apo, which has been called "the most successful marine reserve in the Philippines,"[14] has been the subject of glowing press accounts[15] and is seen as a model for other sites in the developing world. It's true that the Apo Scenario provides a ray of hope in an otherwise rather bleak situation, but there are good reasons not to mistake it for a panacea. The most obvious problem is the long lag between the creation of the reserve and the beginning of the spillover effect. People living on the edge of poverty can't wait nine to thirteen years to see fish catches increase. And unless other changes are made, they're certainly not going to close down a section of their fishery, diminishing their food supply for a decade or more, even if they believe that catches will eventually increase.

For a reserve such as Apo to work, local residents need an alternative source of income while fishing stocks increase (and probably even after). In Apo this was accomplished in part by using the sanctuary as a tourist attraction—bringing in scuba divers and snorkelers to enjoy the island's beautiful reefs. In Apo's case, the money added to the island economy by tourism more than made up for what was lost to fishing.[16]

A far thornier problem is "transplanting" the Apo model, for in some important ways Apo is an unusual site. It is a small island, less than two hundred acres, with only a few hundred inhabitants. Sedimentation from the tiny island isn't a major problem for Apo's reefs, but, as we will see, it is a tremendous impediment for reefs just off mainland areas. And the island's small population doesn't pose the problems of nutrient loading and pollution found in areas with higher population densities. Of course, too few inhabitants can be just as problematic. On reefs without a permanent population nearby, there's no one to enforce fishing bans.

For all the potential problems, the Apo model has inspired the government of the Philippines to attempt other reserves throughout the province. Nineteen such marine reserves have been created to date, and a multimillion-dollar USAID project is currently under way attempting to develop hundreds of Apo-like reserves throughout the country.

Not everyone is convinced that these plans will work, however. "It is messianic for this project to say that it will combat the problem," alleges Filipina Sotto, the head of the Marine Biology Section at the University of San Carlos on the nearby island of Cebu.[17] Sotto points out that there is no shortage of such megaprojects in her country, projects that do more to enrich foreign consultants than to help local people.

"Unless the fisherfolk are empowered in the truest sense and given alternative livelihoods," says Sotto, "then the remaining resources will not be preserved and sustained."

Another scientist is equally doubtful about these programs.

"I lived in the Philippines for six years and speak two dialects," he says, "and the longer I stayed, the more I realized how little I understood about the place and why things happened. I would be amazed to learn that there is even one other place than Apo where fishing access is actually limited. There are far too many fishermen, kids dying left and right, people living on the edge, to make it possible to implement most restrictions except in books and dreams."

And yet others, with just as much field experience, argue that marine reserves are an indispensable part of any plan to preserve coral reefs.

Gary Russ, a marine biologist at Australia's James Cook University, has been studying Apo Island ever since the reserve was created. He considers reserves our last best hope, largely because traditional fishery management has failed all over the world—even in countries with tremendous resources, such as Canada, which allowed the Grand Banks cod fishery to collapse in 1992.

"If the Canadians cannot get it right with their most valuable fishery," he asks, "what hope for fisheries on coral reefs?" The latter are, after all, far more complex and are located in poorer nations.[18] Russ concludes that "given the critical levels of overexploitation of many coral reefs, marine reserves may be the *only* management option available . . ."[19]

Apo may not provide *the* answer, but for coral reefs there is no one right technique. Sotto is correct about the need to empower local fishermen, and about providing alternative livelihoods—and it should be noted that both were done at Apo. Perhaps it may prove anomalous, but Apo is a genuine success story, and in the fight to save coral reefs, such triumphs are simply too rare to ignore.

13

Return to Oceanus

For all at last return to the sea—to Oceanus, the ocean river, like the ever-flowing stream of time, the beginning and the end.

—Rachel Carson, *The Sea Around Us*

I was so intent on locating my prey that I nearly missed the peculiar object hidden among the gently waving blades of sea grass. This was in the mid-1970s. I was hunting stone crabs, whose weighty, blue-tipped claws (washed down with cheap bottles of Mateus) made up a large part of my diet in those languorous Key West days of self-indulgent poverty. I'd dive below the thermocline, the dividing line between the warmer surface waters and the deeper and colder depths, and peer into the eroded limestone caves and hollows. The trick was to hold your breath long enough for your eyes to grow accustomed to the dim light. Grabbing hold of the rock, I'd press my mask up to a cave and wait to detect movement. Sometimes it was a fish. I'd let them pass, since I didn't know which species made good eating or were even edible. But sometimes I could make out the outline of my favorite crustacean there amid the gloom, nervously arranging its giant claws for battle.

I used a wire coathanger in those days, extended into a straight line culminating in a hook. Today it's illegal to use such a device. I'd insert the hanger into the cave until the hook was positioned behind the crab, and then give a tug. Usually the crustacean would come barreling out of its lair, like a bronco released from a rodeo chute. I'd snatch the stone crab before it made its getaway, and thrust it into the nylon bag tied to my wrist. You had to be careful about holding the crabs, since the pincers on the larger ones were fully capable of breaking your fingers—at least that's what the more experienced beach people said. Perhaps they were glamorizing their own exploits. Or they could have been having fun at the expense of the college dropout from Iowa. But the stout claws of the stone crabs looked capable of snapping a human finger, or even severing it. To avoid this, I grabbed them in the only manner sure to immobilize their giant pincers: by the claws themselves.

I considered my existence very romantic, very Hemingway. The great author was after all, Key West's most famous historical resident. In reality, my stone-crab hunts were my own small contribution to the destruction of the marine life of Key West. (Which, in its way, *is* Hemingwayesque, for he left a trail of slaughtered animals behind him wherever he went.) Today I know that you should only take one claw from each crab, since they can regenerate them. Back then, however, in my near-total ignorance of anything concerning life below the waterline, I caught and cooked the crabs whole, eating only the claws and tossing the rest away. A minor abuse. If I had had more money and better equipment, I could have inflicted some real damage to the environment. The irony was that when not needlessly killing stone crabs, I spent much of my time lying on the beach reading philosophical treatises about "getting in touch with nature." Indeed, I felt that I was becoming more and more in tune with the natural world with each crab I killed.

At any rate, I was searching for crabs when something glinted tantalizingly in the sea grass surrounding the rocks. I had once found an Indian-head penny in these waters, and so was always on the lookout for more treasures (piles of gold doubloons from shipwrecked galleons dominated my imaginings). I swam deeper to have a look. Brushing aside the sea grass, I found the object: it was dark and heart-shaped. I grabbed it and headed for the surface.

Sitting on the sand, I examined my "treasure" more closely. It was as black and shiny as obsidian, and felt nearly as hard. It was clearly a seed—two inches across and slightly less than that from base to tip. I had never seen one like it, and neither had any of the old-timers I showed it to on the beach.

"Some kind of seaweed pod," one of them pronounced sagely. "Most likely," he added with an afterthought of humility.

A couple of years later, during my silversmithing period, I made a necklace pendant out of the seed by fastening a silver cap to it, which contrasted nicely with its dark luminescence. The man who bought it for his girlfriend happened to be a graduate student in botany.

"Where did you get the seed?" he asked as I wrapped up his gift. I told him, repeating the part about it probably being from a seaweed.

"It's not," he said. "It's from a tree that grows along the banks on the upper reaches of the Amazon." Later he brought in a book of the flora of Brazil. Sure enough, there was a photograph of the seed, and the tree it grew from.

It is fascinating to reflect on this beautiful seed's journey, from the interior of Brazil to the Florida Keys—a journey of at least two thousand miles. First, of course, it had dropped from the tree into the Amazon, where the current carried it downstream. The Amazon doesn't simply flow into the ocean, however. At times the Atlantic rushes upstream like an impatient lover, at a speed of nearly fourteen miles an hour, producing a wave sixteen feet high, called a *pororoca*.[1] This "tidal bore" gradually slows down, but its effects are still measurable nearly five hundred miles upstream.[2] So it is impossible to say with precision when and where the seed entered the ocean, for the Atlantic and the Amazon are not so neatly separable. This division between the two is further blurred going in the opposite direction, off-shore, where river sediments form a huge, fan-shaped underwater structure known as the Amazon Cone, spilling over four hundred miles to the northeast.[3]

At any rate, rather than sinking into this underwater mountain of sediment, the seed had floated, catching the massive South Equatorial Current as it flowed northwest, mixing with the Guiana Current, and running into the Caribbean Sea. From here the seed bobbed along at a rate of twenty miles a day, past the mangrove coasts of Venezuela and Colombia. Here it could have gotten caught in a gyre, a spiral of water that scours the large cul-de-sac formed by western Colombia, Panama, Costa Rica and Nicaragua. If that had happened, I never would have seen it. But the seed had stayed free of this eddy, and as it reached the western boundaries of the Caribbean, trade winds carried it northward, where it passed through the Yucatán Channel (a stretch of water one hundred miles wide between the Yucatán Peninsula and Cuba) and into the Gulf of Mexico.

Again, the little seed could have been caught by one of the numerous westward currents and swept toward Mexico, where it would have been washed ashore anywhere from the Bay of Campeche in the south all the way up to the barrier islands of Texas in the north. But the main current in the Gulf is the Loop Current, a huge and powerful stream of water that sweeps between Cuba and Florida, where it joins the even more powerful Florida Current, washing past the Keys on its way into the Atlantic, where, farther up the coast, it is renamed the Gulf Stream.

It was here, between Havana and the Florida Keys, that this Amazonian seed was at last plucked out of the main channel by the fingers of an eddy, carried by smaller currents and tides into Hawk Channel, the body of water separating the Florida Reef tract and the islands themselves, and finally came to rest in a bed of sea grass in the shallow waters off Key West.

Far from being unusual, such journeys are commonplace, for ocean waters are forever moving, carrying with them not just the waters from land, but whatever flows into those waters. Sometimes these are living beings, such as my Amazonian seed. Larger creatures have also been known to make these unintended voyages. In 1910 a crocodile from Venezuela's Orinoco River washed up, alive but weary, on the shores of Grenada, more than a hundred miles from South America.[4]

We learn as schoolchildren that all rivers flow to the sea, yet the significance of this truth is rarely understood by those not raised at the water's edge. If oceans are the arteries of the planet, then rivers are the veins, constantly returning water to its source. As a child I once stuffed a message inside a bottle, corked it, and tossed it into a stream in northern Minnesota. I remember the thrill I felt at the time watching the bottle disappear downstream, knowing that there was a chance that a boy living in a Norwegian fishing village might someday read my "GREETINGS FROM AMERICA!"

The problem for corals is that water coming from land contains more than just Amazonian seeds, a few Venezuelan crocodiles, and note-filled bottles from Minnesota. Rivers today carry with them an increasing load of sediments, agricultural chemicals, nutrients, and other pollutants. Each causes its own peculiar problem for coral reefs, and scientists argue over which is more damaging to reefs. On one point there is little disagreement: the combination of these factors is reducing water quality in many places around the world, and this deterioration of water quality is threatening coral reefs. Although those factors are less dramatic than, say, blast fishing, they probably pose a greater threat to coral reefs in the long term.

"Blast fishing causes damage that's similar to that caused by a storm," says coral scientist Evan Edinger. "Coral can recover from a lot of blast fishing. But until you clean up the water, corals are not going to come back. It's the difference between acute and chronic problems."[5]

Some scientists blame the news media for focusing on sensational acute anthropogenic problems—such as ship groundings and oil spills—while ignoring the more important but less spectacular chronic ones.

"[T]he greatest impacts to coral reefs," argued two scientists in a recent paper, "are actually everyday events that are not newsworthy; chronic stresses that have an imperceptible short-term impact, e.g., sewage and agricultural pollution [and] sediment runoff. . . ."[6]

Sewage flowing directly into the sea is a threat to corals, which prefer low-nutrient waters. The broad outline of the problem is well understood. As one coral scientist put it, "We already know that dumping a lot of shit in the water isn't good for much other than the *E. coli*."[7] But the story is a complex one, as scientists studying Kaneohe Bay, Hawaii, found out. The rise, fall, rise, and decline of this one site has been described as a textbook case of the effects of eutrophication, the enrichment of water with nutrients. And so it is, in all its ambiguous splendor.

Kaneohe Bay is well-known to American baby boomers, by sight if not by name. The opening shots for the television series *Gilligan's Island* were filmed there. Kaneohe Bay is the largest embayment in the Hawaiian Islands. It is located on the northeast coast of the island of Oahu, third largest of the eight major islands in the Hawaiian Archipelago, but with nearly four times the population of all the other islands combined. The only true barrier reef in Hawaii lies at the bay's opening. An extensive fringing reef surrounds the bay, and there are several patch reefs, all within easy reach of snorkelers, described in 1915 as luxuriant "coral gardens."[8]

As the capital, Honolulu, grew to the south, the population of the city of Kaneohe likewise swelled, from 5,387 in 1940 to 30,000 by 1960.[9] By 1977, as much as 5 million gallons of treated sewage was emptying directly into Kaneohe Bay daily.[10] Alarming changes in coral cover and the growth of macroalgae were noticed at around the same time. In a two-part series of articles titled "How to Kill a Coral Reef," zoologist Robert Johannes outlined the catastrophic changes occurring at Kaneohe Bay, reporting that "the once-thriving reef communities in the bay's southern basin where the two major sewage outfalls are located have been destroyed; more than 90 percent of the corals are dead."[11]

In many locations throughout Kaneohe Bay, thick sheets of green bubble algae, *Dictyosphaeria cavernosa*, had replaced corals. "The overgrowth of this alga has paralleled the increased sewage inputs and the accompanying increase in nutrient levels in the bay," wrote Johannes.[12]

Public outcry over this deterioration came at a time of heightened environmental awareness, and so in 1977–78 two sewage [illegible] to empty outside of the bay, into the ocean [illegible] The remaining sewage pipe was also di- [illegible]ge in the reef biota following those alterations was astonishing. Algae decreased, and the particle feeders that had replaced corals (zoanthids, sponges, and barnacles) themselves gave way to new corals.[14]

"Reef coral recovery in the bay was documented six years after termination of sewage discharge," pronounced one overview.[15]

Kaneohe Bay was widely hailed as a success story. Nature was resilient; reefs could bounce back from anthropogenic insults. This recovery of coral communities between studies done in 1971 and 1983 was called "remarkable."[16]

But follow-up studies soon had scientists scratching their heads. Between 1983 and 1990, coral abundance failed to increase, and the green bubble algae that had first signaled an "ecosystem under stress" doubled its coverage in that time.[17] What happened?

No one knows for sure. The most complete study of the problem to date concluded that "factors contributing to the health or decline of reefs in Kaneohe Bay are more subtle and complex than the point-source sewage outfalls of the past."[18]

There is no doubt that diverting sewage helped the reefs recover (though what harm the solution is causing in the deeper waters of the surrounding ocean remains unexamined). The problem is that, just as in Jakarta Bay, there are so many other "subtle and complex" factors affecting Kaneohe Bay. Some were already known in 1970, when Johannes wrote his first article on the Kaneohe's decline.

The introduction of intensive, industrialized agriculture on Oahu changed the landscape dramatically. Small fields of taro were replaced by giant farms producing sugarcane, pineapple, and rice for export. By 1930 a series of irrigation ditches rechanneled a large percentage of water that had once flowed into Kaneohe Bay. Fields cleared of existing thick coverage allowed soil to erode and make its way to the bay.

"There are vivid red plumes in the bright blue sea down-current from every stream mouth," wrote one observer.[19]

But agriculture wasn't the only culprit. Urbanization caused problems of its own, aside from sewage. Reported Johannes:

> In the bay watershed, land clearing for subdivisions is done with little regard for soil conservation. Laws requiring the replanting of land cleared of vegetation are not enforced and the bay turns red and opaque several times a year after heavy rains.[20]

In 1969 it was determined that the combination of both of those sources (and probably others as well) had covered the bottom of Kaneohe Bay in places with a layer of sediment nearly five feet deep.[21] Highway construction in the late 1980s added even more sediments to the Bay. One stream emptying into Kaneohe Bay was found to carry six times more sediments after the construction than it had before.[22] Sediments were apparently just as important as excess nutrients in the downward spiral of Kaneohe Bay. So, in the end, reducing sewage-based nutrients didn't "cure" Kaneohe Bay. That's because the problems facing coral reefs are often just as complex and multifaceted as the reefs themselves. As Barbara Brown, the director of England's Centre for Tropical Coastal Management Studies, succinctly put it, "Single stresses are rarely found in the real world."[23]

Ironically, one of the primary threats to the "rain forests of the sea" is the sedimentation caused by the destruction of tropical rain forests themselves. Around the globe, rain forests are being wiped out at an alarming rate, the victim of slash-and-burn agriculture and logging. Just as coral reefs are the treasure troves of marine biodiversity, rain forests are storehouses of terrestrial biodiversity. The richness (and the devastation) of tropical rain forests has been well documented in scientific and popular literature. Thanks to the writings of the ecologist E. O. Wilson, we know, for example, that rain forests, while making up only six percent of the earth's terrestrial surface, are believed to contain half of the total number of species on the planet.[24] In a single hectare (about two and a half acres) of Panamanian rain forest, a researcher estimated that there were eighteen thousand species of beetles—three-quarters the total number of species in all of North America.[25]

As with coral reefs, it's hard to imagine the profusion of life in a tropical rain forest, without actually experiencing it. With this rationale in mind (and the added justification that if coral reefs really are analogous to rain forests, I *needed* to visit the latter for comparison), I played hooky from the International Coral Reef Symposium in Panama and signed up for a day trip to the Barro Colorado Nature Monument, the oldest protected tropical rain forest in the world.

❄

It is five-thirty on a June morning, and six of us are trying to stay awake as our microbus bounces merrily down the bumpy highway that parallels the Panama Canal as it heads northwest to the Atlantic Ocean. Forty-five minutes later, and halfway across the isthmus, the bus skids to a stop. We stumble out and are promptly herded onto a small boat, which heads off into Lake Gatún, making for Barro Colorado Island.

Our guide is a Costa Rican ornithologist named Mario. He is tall, with one of the squarest jaws I've seen outside of comic books, and a thick mustache. He wears a fedora-style rain hat, à la Indiana Jones. In fact, Mario looks strikingly like a Latin American version of that famous film adventurer as he stands near the bow, one foot placed cavalierly on the gunnel, a muscular hand resting on his hip, surveying the water's edge through binoculars. All that's missing is the coiled whip hanging from his belt. When he is satisfied that we are safely under way, Mario turns to greet us with a hearty "Good morning, everyone." It's still early, and we respond in a desultory fashion, like a group of shy schoolchildren on a field trip. Mario directs a toothy smile back at us and shakes his head. Although he is too polite to say it, you can almost see the word as it runs through his mind: *gringos.*

Few visitors are allowed on Barro Colorado Island itself, which is run by the Smithsonian Tropical Research Institute. But as the boat glides quietly into one of the many coves surrounding the island, tall trees draped with dark vines and silvery epiphytes (plants growing on other plants) draw over us like an immense green cloak. We spot several toucans perched on tree limbs, their enormous, multicolored beaks instantly drawing the eye. Mario explains that toucans use those pretty beaks to steal eggs from other birds' nests.

"What?" he jokes. "You thought they just ate Fruit Loops?"

Turtles bask on rocks near the shore, surrounded by swarms of brilliant blue butterflies, busy licking the salt from the reptiles' eyes. High above, a troop of howler monkeys spots us, and true to their name, they begin to howl, an arresting sound that has been likened to the roar of "an enraged jaguar."[26] A sweet-faced female with a baby clutched to her breast gazes down at us. The infant pulls away from the teat to regard us with its enormous eyes. What does it see when it sees us: distant relatives drifting on the waters below, cooing excitedly? Are we threats? Curiosities? Suddenly a young howler swings to the fore, practicing his warning cry on us. His adolescent voice breaks as he heaps down abuse. This goes on for a while until finally some older monkeys in his troop bark at him in a manner clearly meant to convey that, Yes, we've seen them, so for God's sake, shut up! He stops, but continues to glare sulkily at us as our boat passes. As we drift around a corner, I can still see the young monkey mouthing silent curses in our wake.

We head over to the Gigante Research Station, on the mainland just south of Barro Colorado, but still part of the rain-forest nature monument. After docking, we climb off the boat and begin our trek into the rain forest proper. After reading about the high biodiversity, I am surprised by the relative lack of animal life around. Where are all those thousands of beetles? But, as Mario explains, 70 percent of the animal species in the forest live a hundred feet above us in the forest canopy. The many layers of leaves above us are so thick that less than 3 percent of the available light (the total spilling down onto the top) reaches the ground. Still, there are leafcutter ants everywhere, forming long lines as they carry bits of leaves back to their underground nests, where they will be chewed to a soft pulp and then stored as a symbiotic fungus grows on them. It is this fungus that the leafcutters eat, not the leaves themselves, although their "farming" may be as important to the nutrient cycling of the rain forest as damselfish farming is to the coral reef. The ants consume 250 pounds of leaves per acre in a year, an amount equal to the total eaten by all vertebrates found in the rain forest.[27] Fortunately, we don't happen upon the leafcutters' relatives, the *Paraponera clavata*, which, at an inch long, is the largest ant in Central America, and the most dangerous: its painful sting can burn for several days.

The name *Panamá* is thought to be an Indian word meaning "land of abundant butterflies,"[28] and it's easy to see why. Butterflies are everywhere, some the size of postage stamps, others as big and brightly colored as costume-ball masks. In movement, color, and sheer profusion, butterflies suggest the fishes on coral reefs. Their brilliant colors distract predators with a technique dubbed "flash and dazzle,"[29] just like reef fishes. And many butterflies and moths have false eye spots similar to those found on reef fishes.

More monkeys swing through canopy above. Spider monkeys, long and slender with dark fur, go swooshing by, arm over arm, hardly taking the time to chatter at us. A little farther on, a troop of white-faced capuchin monkeys peer down at us and then they scurry off, too. A pair of large *Ameiva* lizards are intertwined, mating on the path directly ahead of us. Not wanting to interrupt their intimate moment, we stand around for a while. I might feel like a voyeur, except that there is absolutely no visible action occurring. It's like looking at a photograph of lizards copulating. After five minutes of this, we grow impatient and decide to step over them. As we do so, the pair disengages and hurries off into the undergrowth.

Because Mario is an ornithologist, he's constantly pointing out birds: parrots, a greater ani, several species of hummingbirds, a rare crested guan. He usually identifies the birds by their calls, and we follow him as he searches for the singer, returning the call. At one point we stop along the path while Mario softly whistles an excellent imitation of the bird he's trying to attract.

The bird trills a few notes from a distance.

Mario replies in kind.

The bird chirps again, this time from closer by.

Mario is about to repeat the short birdsong, when a young man in our group points to a branch by my left shoulder and says softly, "Mario, what's that?"

I freeze. Mario looks over and his face lights up.

"It's a poison-arrow frog. Grab him!"

I turn slowly and there it is: a tiny frog at eye level. It is beautiful—jet black with bright greenish yellow spots, about the size of my thumb. Without thinking, I do as instructed. The frog doesn't even attempt to get away as I scoop it up. Clutched in my right hand, it pulsates like a tiny heart. The others in the group gather around to see. He's so small and insubstantial I'm afraid I'll hurt him as he begins struggling to escape.

While everyone is admiring him, Mario tells us about how this type of frog gets its name. When scared, its skin produces a toxin so poisonous that Indians wipe their arrow tips with the secretion.

"That way, even if they only nick their prey, it still dies," he says with admiration.

Silence. Then the same man who first spotted the frog asks, "Wouldn't the frog be secreting those toxins right now?" His use of the conditional mood makes me queasy.

"Sure," chirps Mario.

I set the little frog down and he hops away.

Mario laughs.

"Just don't lick your hand before you can wash," he says. "Okay, let's see if we can find that bird." He turns and starts whistling again.

The rest of our group stares silently at me for a moment. They glance at my right hand and smile nervously, before heading—quickly, it seems—down the trail.

"Heh, heh," offers the young man who first pointed out the frog as he walks past.

I examine my right hand for open cuts, and am relieved to find none. I imagine my wife standing over my coffin, keening, "What possessed him to pick up something called a 'poison-arrow frog' in the first place?" There's nothing to do about it now, however, so I continue down the trail looking for birds and other life. But for the remainder of our hike, a portion of my brain is constantly tracking the whereabouts of my right hand and its proximity to my mouth, and periodically issuing warnings to the limb: "Stay away from the mouth. Stay away from the mouth."

Of course, the real profusion of life here, the bulk of the biomass, is not in the poison-arrow frogs or the birds or the monkeys or the insects, but in the plant life, the primary producers that make all the rest of it possible. The tremendous amount of solar energy available to tropical rain forests (approximately 30 percent more energy per square foot than falls on a backyard in Columbus, Ohio in a year[30]), the abundant rainfall (102 inches a year at Barro Colorado, compared to a measly 37 inches for central Ohio[31]) and the year-round growing season combine to make these biomes tremendously productive. Scientists estimate that tropical rain forests produce nearly four pounds of dry plant matter per square yard annually—compared to just over two pounds for the typical deciduous temperate forests.[32] (Ecologists estimate the primary productivity of coral reefs as ranging from between three and thirteen pounds per square yard annually.[33])

Although a great number of plant species grow in the rain forest, that fact is not apparent to the casual observer. Many trees and plants are variations on some common themes, evolutionary adaptations to their environment: pointy leaves, buttressing roots, showy flowers, a flattened, umbrellalike crown. The most striking contrasts aren't between individual species so much as between occupants of different zones, with plants growing at different heights tending to look like their neighbors. Zonation is another property shared by coral reefs and rain forests, with several overlapping zones identified in the forests, from the scrubby understory to the canopy layer, over a hundred feet above. Each zone is defined by its microclimate: the amount of sunlight reaching it, temperature, relative humidity, and wind velocity, as well as other factors. For example, when the temperature is nearly 83°F in the canopy, it's 80°F at the forest floor. That might not sound like much, but an average difference of three degrees can radically alter which plants can survive and which cannot. The difference in relative humidity is even more pronounced. When the canopy is a muggy 86 percent relative humidity, the floor is a sweltering 94 percent.[34]

This tremendous biomass of plant life would seem to suggest rich soils, but the opposite is the rule in tropical rain forests. One secret to the profusion of plants is the extremely fast and efficient recycling of rain-forest nutrients. The life of a temperate-zone forest is . . . temperate. That is to say, slow. In the higher latitudes of slanting light and frigid winters, leaves fall to the ground and slowly, over the course of a few years, decompose into a thick layer of nutrient-rich humus. But life in the tropics is intemperate, impatient, hasty. The fallen leaf is set upon by a host of voracious decomposers the minute it hits the ground. In less than a year, nearly every molecule has been incorporated into another being[35]—a process known to ecologists as "tight" nutrient cycling. One result of this phenomenon is that there are very few organic nutrients in the soil on Barro Colorado and in the rain forests nearby. Of course, this doesn't bother the plants here; they're well adapted to this situation, with roots growing only a few inches deep.

The problem comes when rain forests are destroyed. Then tropical downpours pummel the barren land, soils without humus to soak up the water or even slow its flow. Sediments pour into streams and rivers. Soil erosion in pristine forests has been calculated at anywhere between four and forty-four pounds per acre annually, given a variety of factors. But once a forest has been cleared, the exposed land erodes with astonishing rapidity. In land converted to agricultural use in Asia, it isn't unusual to find yearly erosion rates measuring in the *tons* per acre.[36]

The Panama Canal basin, of which Barro Colorado is a part, is a good example of the problem of deforestation. In 1952, 83 percent of the 724,000-acre area was covered by rain forests. By 1978 the forest had been reduced to just 232,000 acres—32 percent of the basin.[37] The destruction of the rain forest is, of course, a worldwide calamity. By 1979, just 56 percent of the prehistoric tropical rain forests around the globe remained.[38] Despite mounting scientific evidence of the ecological consequences of the practice, the 1980s saw a frenzy of deforestation—with the result that by 1989 less than half the prehistoric cover remained.[39]

Like rows of dominoes, coral reefs and the fisheries they support and the people who depend on them for food and livelihoods are all in danger of toppling as the rain forests collapse. Corals can withstand surprisingly high sediment loads *if* the situation is temporary. This was proven rather graphically in 1980 after a freighter ran aground on reefs in Hawaii and its cargo of 2,200 tons of powdered clay was tossed overboard in order to refloat the vessel. Despite fears that the sediment would kill huge amounts of corals, only nearby corals actually buried beneath the clay were destroyed.[40]

What corals can't survive, however, is *chronic* sedimentation: constant elevated doses of terrestrial matter.

Caribbean reefs have been damaged by increased sedimentation (including the Panamanian reefs of the San Blas Islands[41]), but this is a worldwide problem, documented in at least fifty countries.[42] Most alarming, the problem is worst in the heart of coral biodiversity. Half of all sediments emptying into the global ocean enter from rivers and streams in this area (which includes Papua New Guinea, the Philippines, and Indonesia). Another 25 to 30 percent of land-based sediments comes from other areas of Southeast Asia,[43] many of which include coral reefs.

It's fitting, then, that the classic study of the downstream effects (literally) of logging and sedimentation on corals comes from this part of the world. The 1988 paper by Gregor Hodgson and John Dixon is also important because it was one of the first attempts to take an integrated approach, "involv[ing] investigations of links among ecosystems—terrestrial, riparian [along rivers], and marine—and includ[ing] both socioeconomic and ecological perspectives."[44] Not to put too fine a point on it, what they're talking about is braiding.

Hodgson and Dixon studied the effects of logging in the rain forests of Palawan, a narrow, 260-mile-long island separating the South China Sea from the southernmost islands of the Philippines. Just two decades ago, Palawan was considered one of the last great stretches of pristine land in the archipelago, an Edenlike territory of spectacular tropical mountain forests, filled with endemic species, such as the Palawan bear cat and the Palawan mongoose and some three hundred kinds of butterflies, surrounded by vast, uncharted coral reefs and blessed with teeming fisheries of yellowfin and skipjack tuna.

Then, in a pattern that is depressingly familiar, depletion of resources elsewhere (combined with civil strife in adjacent areas) increased both population and industrial pressures on Palawan. Throughout the virgin forests of Palawan, the primordial sounds of insects buzzing and birds calling were quickly replaced by the modern din of chain saws and heavy diesel-powered equipment. The forest that covered 92 percent of the island in 1968 remained on just 50 percent of it by 1987.[45]

E. O. Wilson has coined the term "centinelan extinction" to describe the anonymous loss of rare species when pristine areas of high biodiversity are wiped out in a short period.[46] Almost certainly the Palawan forests experienced a number of centinelan extinctions. In addition, Hodgson and Dixon found, there were other—economically quantifiable—consequences as well. It's important to document these changes, for, as the authors of the study point out, "Government policymakers are familiar with the language of economics and often make decisions based in part on predicated economic returns, not [on] abstract natural science concepts such as the ecological value of a 'keystone species.'"[47]

The authors looked at logging activities within a remote thirty-square-mile region on the northern tip of Palawan, and its effects on reefs and fisheries in an adjacent (and slightly larger) marine area named Bacuit Bay. Building on recent studies of erosion caused by logging, the authors determined that the greatest erosion wasn't caused by logging itself, but by the roads built to bring equipment in and take felled trees out. Though roads accounted for only 3 percent of the land area, they caused 84 percent of surface erosion.[48]

Hodgson and Dixon determined that logging in the region increased sediment deposition in Bacuit Bay at levels up to sixty times higher than the normal monthly rate.[49] As a result, the coral reef closest to the mouth of the Manlang River (which drained most of the study area), lost nearly half of its coral cover.[50] They also found that for each 1-percent annual decline in coral cover, fish biomass decreased 2.4 percent.[51]

Finally, the pair looked at the downstream economic effects of two options: continuing the logging operations according to plan, or banning logging in the region. They made a startling discovery: by banning logging, total revenues would more than double.[52] The distribution of income is important as well. Logging gives the greatest economic benefits to fewer people than the alternatives of fishing and tourism,[53] with most of the money going to wealthy business owners outside of the logged region.

The good news is that Filipino government planners seized on the economic analysis provided by the study and banned logging in the area, enforced regulations against slash-and-burn agriculture, and created a marine reserve in Bacuit Bay. After revisiting the area in 1996, Gregor Hodgson observed that new vegetation had covered the land scarred by logging and poor agricultural practices. "The coral reefs previously damaged by sedimentation in 1996 appear to have recovered nicely," he reported.[54]

Now for the bad news: Hodgson found that the fisheries of Bacuit Bay have actually declined—despite the ban on logging, despite the resulting reduction in sediments flowing into the bay, despite the designation as a marine reserve, despite the increase in coral cover. "Decimated" was the word he used to describe what has happened to the marine life of the bay. The reason is overfishing. As fisheries were depleted elsewhere, impoverished fishermen heard about the good conditions in Bacuit Bay and moved there. Between 1980 and 1995 the population of the bay area doubled—and local consumption of fish rose with it. That's just part of the story, however. Like a character in a Greek tragedy, Bacuit Bay appeared fated to be thrust into the larger market economy—if not through logging, then through fishing. One day an exporter set up business in a village on the bay. Soon he was shipping as much as twenty-two tons of iced local fish over to Manila each month.[55]

There is a veritable smorgasbord of lessons to choose from concerning what happened in Kaneohe Bay in Hawaii and Bacuit Bay in the Philippines. The most obvious lesson is that when humans add too many nutrients or too much sediment to tropical coastal waters, corals suffer. The point of origin for these damaging substances can be near to or distant from the coasts themselves; it makes no difference. As Rachel Carson reminds us, all things return to the ocean. For optimists there is this lesson of hope: When institutions take action to reduce nutrification and sedimentation, corals can recover from previous decline.

But pessimists can pick lessons more suited to their bitter palates. Preserving reefs is clearly not a relatively simple matter of banning logging in a section of rain forest or diverting sewage from a single location, for without a healthy population of reef fish in Bacuit Bay, it is probably only a matter of time until the corals there give way to algae. Coral reefs are complex, and so are the problems facing them. Our solutions must be just as intricate.

"It's not a matter of picking and choosing which problem we attack," says marine biologist Thomas Goreau. "We have to do it all; all at the same time."[56]

14

Disasters, Catastrophes, and Tragedies

For perhaps the first time in the history of our planet, a single species may be close to exerting an effect that rivals the importance of . . . natural controls.

—Dennis Hubbard, coral scientist

When is an event a disaster? When is it a catastrophe? When a tragedy?

The three words are often used interchangeably. But to understand the threats to coral reefs, and to find solutions to reef degradation, it's important to define these terms with some precision.

Start with a simple and famous example: Krakatau, the volcanic island that blew itself nearly to bits on the morning of August 27, 1883. For anything that lived on the island, located between Sumatra and Java, the tremendous explosion (estimated to have been the equivalent of 100 to 150 megatons of TNT) was a cataclysm of the highest order. And there was a lot of life present on Krakatau. The volcano's slopes were covered with all the lush biological extravagances of a tropical rain forest: innumerable butterflies, ancient trees connected by looping lianas, insects by the millions scurrying around the canopy, fantastic birds by the thousands.

After the explosion, nothing.

At least at first. Nearly a year afterwards, a French expedition landed on what remained of Krakatau's southern extremity, renamed Rakata. After spending hours climbing over the rubble, the expedition's naturalist reported that the only life-form found was a single microscopic spider: "this strange pioneer of the renovation was busy spinning its web."[1]

Soon grass was spotted growing on the island. By 1928 Rakata was covered with nearly three hundred species of plants. Crabs and insects, birds and rats, all made a comeback. Today, while the species in residence on Rakata remain in a state of flux, as E. O. Wilson has put it, "a century after the apocalypse, [it] gives the impression of life on a typical small Indonesian island."[2]

Let's not consider the effects of the eruption beyond the island itself, where tidal waves killed tens of thousands of people. Limiting our view to Krakatau, then, we have an example of a disaster: a natural event from which life can recover.

Disasters of varying intensities are quite common in nature. In fact, though they are devastating to individuals and often to whole communities, they can be beneficial to an ecosystem's health over the long haul. This counterintuitive way of thinking is rather new to science. For a long time ecologists believed that the ecosystems with the highest species diversity were in a state of near equilibrium—arranged just so, with each creature occupying its perfect niche, if not in perpetuity, then at least over very long stretches of time. That belief changed in 1978, after an American biologist named Joseph Connell published a paper on the "intermediate disturbance hypothesis."[3] Connell proposed that high diversity actually *depends* upon periodic disasters or, as he called them, disturbances. Such disturbances experienced by coral reefs (in the form of hurricanes, for example) and tropical rain forests (windstorms, insect plagues) prevent a few dominant species from pushing "inferior" ones out. Without the occasional cataclysm, diverse ecosystems would actually lose species richness.

"It's having that little disturbance now and then that is the spice of life," explains coral physiologist Erich Mueller.[4]

According to Connell's hypothesis, which is now widely accepted, ecosystems such as coral reefs have evolved as they are, not in spite of periodic disasters, but *because* of them.

Granted, Krakatau was more than an intermediate disturbance; it was, by any definition, an extreme disturbance. But that is its value as an example: despite the initial devastation, a century later, life has returned in nearly identical form.

But what if it hadn't? True catastrophes, disasters so formidable that they forever change the face of life on the planet, or at least on a section of it, are also part of a long-established process, one that has taken place without human involvement. Witness the extinction of the dinosaurs—or the Permian crash, which wiped out a majority of marine species 245 million years ago. These were beyond disasters of even a higher magnitude; they truly deserve the name *catastrophe*. But while we may mourn the loss of dinosaurs and of the thousands of other groups of plants and animals wiped out periodically, those catastrophes are not tragedies. To my way of thinking, at least, the term *tragedy* should be reserved for catastrophes caused by humans.

This neat division remains tidy only on paper, however. In the real world, how do we know what role humans have played in a given disaster or catastrophe? It's easy in the case of the Permian crash and the extinction of the dinosaurs: both occurred before mankind's arrival on the scene. Humans were around, but no one believes we played a role in the destruction of Krakatau. Go much further than these examples, however, and the lines between modern disasters, catastrophes, and tragedies get blurred very quickly. And yet it is important to attempt to distinguish among them, for humans are a species in some ways unlike any other that has come before. We are, for starters, superpredators. Our annihilation of the passenger pigeon is a well-known example of a human-induced extinction, but few appreciate the magnitude of that tragedy. Only a few decades before the last passenger pigeon died in a Cincinnati zoo in 1914, her species was the single most numerous bird in North America—accounting for one out of every five birds on the continent.[5] Our abilities as a superpredator may soon push several species of coral reef fishes over the same biological brink that claimed the passenger pigeon.

It is not our tremendous hunting capability, however, that poses the greatest danger to other species and entire ecosystems. We inflict our worst damage inadvertently, casually, often not even recognizing the catastrophic effects of our actions. That's what happened in the case of the Caribbean long-spined black sea urchin. Among marine biologists, the story of its extirpation has become emblematic of this kind of unwitting tragedy.

❊

For centuries, humans in the New World regarded *Diadema antillarum* with the kind of fear and loathing usually reserved for malaria-carrying mosquitoes and man-eating sharks. And with good reason: the echinoderm's hollow spines—each one as thin and sharp as a hypodermic needle—can puncture shoe leather almost as easily as human skin. It wasn't just that they were dangerous, but that they were *everywhere*, often in remarkably high densities. On many reefs it wasn't unusual to find dozens of the creatures packed into a single square yard.[6]

A naturalist visiting the Caribbean in the late seventeenth century warned his readers of the "poisonous" sea eggs (as they were known then) "set about on every hand with prickles."[7]

"No one goes bathing or into the water for any purpose in this region without being warned against the danger of being wounded by the cruel black spines of this ubiquitous sea-urchin," wrote another in 1919.[8] One writer termed the sea urchin "dreaded,"[9] while a modern field book called it "fear inspiring."[10]

The recent authoritative field guide *Reef Creature Identification* reinforces the urchin's frightening reputation, pointing out that "spines easily puncture the skin and break off in the flesh, causing a painful wound. The embedded spines give off a purple dye, causing a slight discoloration under the skin. . . . Victims with numerous wounds may need treatment for shock."[11]

Even the animal's common name, *sea urchin*, is tainted with evil intentions. The word *urchin* comes from the old name for the English hedgehog, which was thought to be a mischievous spirit in disguise.[12] The urchin's behavior only adds to its aura of malevolence. Hold your hand a few inches away from the urchin and watch as the spines pivot slowly toward it, like a battery of surface-to-air missiles preparing for launch.

Given the fearsome capabilities of these creatures, it's not too surprising that when reports first surfaced that a great many *Di-*

adema had died at Punta Galeta, Panama, near the Caribbean side of the great canal, in mid-January 1983, the most common reaction was "Good riddance."

Ecologists, however, were alarmed. Haris Lessios, a biologist with the Smithsonian Tropical Research Station in Panama, watched as whatever was killing the *Diadema* began to spread from Panama and out into the Caribbean. His 1984 *Science* magazine article reads like the notebook of a detective stalking a serial killer:[13] By April, *Diadema* were dying in the San Blas Islands, east of the canal. The path of destruction spread swiftly, and then split in two, simultaneously heading east and west. In early June the killer had reached the Cayman Islands to the northwest. By the end of the month, *Diadema* were dying in both Colombia (to the east) and Costa Rica (to the west).

Over the next thirteen months, Lessios and his colleagues tracked what they concluded was a species-specific pathogen, probably some form of bacteria, which killed its host in a matter of days or a few weeks in a particularly gruesome manner: First, the outer layer of skin covering the spines sloughed off.[14] Next, the spines and other parts lost their pigment and spines began breaking off the central shell, called a *test*. As the disease progressed, the behavior of infected *Diadema* became bizarre. The echinoderms typically graze around patch reefs during the night and return to the safety of the reef at first light. Infected urchins wandered about in the open during the day, becoming easy prey for fish. Those not eaten soon lost control of their tube feet, used to cling to the seafloor. Finally the dreaded *Diadema* literally fell apart—skin and spines and tests disintegrating.

From the path of the disease, it seemed likely that the unknown pathogen was riding surface currents, infecting and wiping out *Diadema* populations as it went. In a little over a year, the trail of destruction had circled the entire Caribbean, traveling nearly four thousand[15] miles and destroying approximately 98 percent of the sea urchin population[16] in an area covering more than a million square miles.[17] Billions of individuals perished. It was the worst mass die-off recorded for a marine animal in modern times.[18]

What happens to an ecosystem when a single living component is suddenly removed? No one knew for sure. That's because any ecosystem, and particularly one as complex as a coral reef, is the biological equivalent of a Rube Goldberg machine. Goldberg was an American cartoonist in the first half of this century, famous for his single-panel drawings of enormously complex machinery for producing spectacularly simple effects. For example, a marble was dropped into a small basket in the upper left part of the cartoon. The basket, arranged on a central pivot, fell, and on the other side of the teeter-totter-like contraption, a nail rose and punctured a balloon. The sound of the bursting balloon scared a mouse, which scampered down a runway and through a turnstile, which, in turn, struck a match. The heat from the match boiled the water in a thimble-sized steam engine, which pulled a miniature train down a short length of track . . . you get the idea. Each cartoon contained several equally bizarre steps, until, at last, a final task was accomplished: a lightbulb was turned on by a frightened beetle grabbing at the pull chain, or a pat of butter was cut neatly by a plummeting knife. Goldberg's machines were a comedic meditation on cause and effect, on the distance and the often unexpected pathways traversed between the two.

Ecosystems are like that. Scientists are continually discovering the quirky and circuitous pathways linking one creature or community to another. Usually this is done under controlled conditions in laboratories, or in small-scale experiments out on a reef. The 1983 mass mortality of a single sea urchin species throughout the Caribbean (an event referred to by biologists as the "*Diadema* die-off") presented scientists with a unique opportunity to observe a huge living Rube Goldberg machine at work. At one end of the machine *Diadema antillarum* had been plucked out. What would happen at the other end?

The answer was not a happy one for coral reefs.

Diadema may have been a dangerous nuisance to humans, but it was the mortal enemy of algae. The urchin consumed so much of the simple plant that the marine biologist Alina Szmant called *Diadema* "the lawnmowers of the reef."[19] With these lawnmowers removed, algae growth went wild, smothering entire reefs. A notable example occurred in Jamaica, where 99 percent of the urchins were wiped out.[20] Coral cover along Jamaican fore-reefs declined from 52 percent before the die-off to just 3 percent in the years following the epidemic. The worst of times for corals was the best of times for its competitor, algae. Algae cover for the same years and locations skyrocketed from 4 percent to 92 percent.

"The scale of damage to Jamaican reefs is enormous," concluded Australian marine biologist Terence Hughes. "A striking phase shift has occurred from a coral-dominated to an algal-dominated system."[21]

The transformation of a biodiverse coral reef into a clump of algae-covered rocks was clearly a catastrophe, but it was also a tragedy.

Decades before the epidemic wiped out sea urchins, the die was already cast due to human activity. To understand the phase shift to algal reefs, we have to enlarge the time frame of our Rube Goldberg machine to include the events preceding the mass mortality of *Diadema.* Long before that catastrophe, another tragedy was brewing. As the population of Jamaica grew from 40,000 inhabitants in 1688 to 300,000 in 1793, the fishery surrounding that island was gradually depleted.[22] In 1962 there were nearly 1.7 million people living on the island. That same year, 24 million pounds of fish were caught in Jamaican waters. In 1968, with a population of nearly 2 million people, the harvest of local fish was cut in half, to 12 million pounds.[23]

Fish have two major effects on *Diadema*—one direct, the other indirect. The direct effect is that some common reef fish, such as the queen triggerfish, eat the prickly urchins. The indirect effect is through competition for food (algae) by herbivorous fish. Both effects came into play as the Jamaican reefs were overfished. Thanks to humans, *Diadema* gradually found itself relatively free from both predation and competition. Its population exploded.[24]

Jeremy Jackson disagrees, at least in part. The marine biologist believes that the Rube Goldberg machine drawn above lacks an important cog.

"*Diadema* abundance is at best a secondary consequence of the degradation of Caribbean reefs," Jackson wrote.[25] The missing cog in the machine, according to Jackson, is *Chelonia mydas*, the green sea turtle, and the reason no one thought of it is that humans had removed the marine reptile from the ecosystem hundreds of years ago.

Jackson points out that in the fifteenth-century, Caribbean green sea turtles were so abundant that a migrating stream of them forced Columbus to stop his ships for a full day to let them pass.[26] Contemporaries of the explorer noted that along the coast of Central America it looked as if you could walk from your ship to land a mile off, merely by stepping on the backs of the multitude of sea turtles congregating there.[27] What the bison was to the North American continent, *Chelonia mydas* was to the Caribbean. And it suffered similar consequences.

Because green turtles were ubiquitous and easy to catch, and contained a large amount of food per individual, they soon became the most important source of animal protein in the diet of the newest residents of the New World. Jackson estimates that there could have been more than 600 million green sea turtles in the Caribbean before the slaughter began.[28] By the end of the eighteenth century, the great masses of green sea turtles had already been wiped out.[29] Today only a few hundred thousand remain in the Caribbean.

Many ecosystems have "redundant" species built in—that is, several species capable of filling the same ecological niche. "The more species capable of filling a role," explained one scientist, "the more resilient will be reef processes to species loss."[30]

But nature creates only a finite number of redundancies. It's possible that all appeared well on the reefs for decades largely because the effect of removing turtles was masked by an increase in algae-eating fish. When those herbivorous fishes were decimated, this too was masked by an increase in the number of algae-eating urchins. When the mysterious pathogen swept through the Caribbean in 1983, however, we had run through nature's backup systems and the coral reef paid the price. But long before that, humans had begun to unravel the braid that held the Jamaican reefs together.

❄

Two giant oceans cover our planet, one made of water, the other of gas. That second "ocean" is the atmosphere. We are so completely immersed in it that we don't think about the atmosphere in those terms, any more than a fish probably thinks about water. And yet the atmosphere is just as much a churning reservoir of fluids as the watery ocean it covers. The two oceans are also intimately related. There is a "dialogue between wind and sea"[31] that gives rise to waves, storms, currents, and clouds that shape the daily lives of both terrestrial and marine creatures. Even air temperature and rainfall over distant landmasses are affected by the great ocean currents. So important is this partnership that meteorologists and oceanographers now talk about these twin oceans as "a single system of two fluids interacting with each other."[32]

One of the most dramatic examples of this relationship is the Indonesian Low, a convergence of oceanic and atmospheric conditions that has been called "the heat engine that drives the earth's climate and temperature."[33] Equatorial trade winds in the Pacific drive sun-warmed surface waters westward, where they form a huge pool of warm water over three hundred feet deep in the western Pacific. So much water is piled up in the western Pacific (some forty times the total volume of water contained in the American Great Lakes) that it actually creates a bulge in the ocean; "sea level" is normally a foot and a half higher in Indonesia than across the Pacific in Ecuador, and the water in this "warm pool" is 14°F warmer. As this massive body of warm water evaporates it creates a tremendous low pressure system: the Indonesian Low. The warm, moist air rises thousands of feet into the sky, where it encounters cooler temperatures and forms giant cumulonimbus clouds that eventually release the water as rain. Meanwhile, back in the eastern Pacific, colder, nutrient-rich water from great depths is drawn to the surface to replace the warm waters blown west by the trade winds.

At least that's how things usually work. But every few years something very different happens. The South Pacific trade winds stall. The pressure at the Indonesian Low rises; in the eastern Pacific it falls. Instead of heading west, warm Pacific water surges six thousand miles east along the equator, where it runs into the coast of South America, splits, and flows both south and north.

This condition, which is apparently cyclical—but not predictable with our current understanding—is known as an ENSO event, an acronym that combines the names of its oceanic and atmospheric components. "El Niño" (EN) was what nineteenth-century Peruvian fishermen called the current of slightly warmer waters that flowed through their fishing grounds each year around Christmas time. Gradually, however, the term has come to refer to the less frequent but more devastating arrival of the much warmer waters in the Eastern Pacific. SO stands for Southern Oscillation, the dramatic seesawing of air pressure between the eastern and western Pacific now associated with El Niño. (Although ENSO is the preferred designation, sometimes the older "El Niño" is used to refer to the phenomenon.)

Scientists are just beginning to understand the profound effects an ENSO event can have on land and sea. It can halt the normal upwelling of nutrients from the cold waters off South America, devastating entire fisheries for thousands of miles. By distorting the flow of the jet stream, an ENSO event can produce milder winters in the higher latitudes of western North America, and increase rainfall as far away as Florida. To the tropical west, in Indonesia, Australia, and India, lands accustomed to daily monsoons can go cloudless for weeks. Plants shrivel, brush-fires sweep through. The parched earth cracks, and soil blows away in clouds. ENSO events are clearly one of the more decisive (not to mention destructive) phenomena produced by the dialogue between ocean and atmosphere. But it wasn't until 1983 that their effects on coral reefs were first suspected.

Residents of Christmas Island in the mid-Pacific first noticed something odd in the summer of 1982 when the sea level surrounding their island jumped by several inches in just a few weeks.[34] They didn't know it, but this was the bulge of the "warm pool" heading east. Then island residents found an alarming number of baby birds dead or starving in their nests. The adult birds that usually provided for them had disappeared.[35] (It was later determined that the adults were far away over the ocean, searching for food.) The strange occurrences had been preceded by a general slackening of the east-to-west trade winds in May. Over the next several months the massive warm pool of western Pacific water headed east, raising surface temperatures and the level of the sea as much as a foot as it went. By October the leading edge of the mass of warm water swept past the Galápagos Islands and then crashed into the shores of South America. Water temperatures rose from their normal levels in the low seventies (Fahrenheit) to the upper eighties.[36] From March 1983 to June of that year, the abnormally warm waters spread throughout a huge area from Central America to the Galápagos Islands, nearly a thousand miles away.[37]

What was occurring, it was later determined, was the most powerful ENSO event of the past two hundred years.[38] Nearly one hundred inches of rain fell on the normally arid coastal plains of Ecuador and Peru, causing populations of grasshoppers and malaria-carrying mosquitoes to explode.[39] Typhoons smashed into Hawaii and Tahiti—neither place having been known for such powerful storms.[40] Droughts ravaged Australia and Indonesia.

On March 18, 1983, at the same time as these catastrophes were occurring, Peter Glynn noticed something else, strange and alarming, going on in the waters surrounding Uva Island, one of a sprinkling of tropical islands in the sheltered Gulf of Chiriquí, on the Pacific side of Panama. Glynn, a marine biologist at the Smithsonian Tropical Research Institute, saw that large areas of coral cover, some over one thousand square feet, had turned ghostly white.[41] Normally these corals ranged in color from iridescent gold to pale green. As Glynn toured the surrounding islands he observed many more such areas, eventually estimating that between 50 percent and 80 percent of the total coral cover was affected.[42] On closer inspection, Glynn found that the coral polyps were intact. Their colors, however, were being slowly sucked out of them, leaving behind large patches of reefs that glowed like the dry bones in a Georgia O'Keeffe painting. And in fact it was the pure white coral skeletons that were visible through the transparent coral tissues.

What was happening was the unweaving of the most basic of coral braids: the one linking coral polyps with their symbiotic algae, the zooxanthellae. Trillions of zoox living within billions of coral polyps were leaving, bailing out into the ocean waters in a process known as coral bleaching. Some of the bleached corals survived, but many others died—deprived of the nutrients supplied by their symbionts.

Even today, bleaching remains by and large a mystery. The title of a 1997 scientific journal article confesses that "Mechanisms of bleaching are poorly understood."[43] Back in early 1983, when Glynn first noticed the mass bleaching off the coast of Panama, even less was known about the phenomenon. It had been observed before, but rarely, and then usually as isolated incidents. The first reports of large-scale mass bleaching anywhere dated from 1979,[44] when corals lost their zoox in a few spots in the western Pacific and the Caribbean.[45]

As Glynn and others watched, the 1982–83 bleaching event spread across several reefs in the eastern Pacific, including in the Galápagos Islands, where 97 percent of hard corals died.[46] Then in June came another unexpected development: corals on the Caribbean side of Panama also started bleaching and dying.[47] Some scientists theorized that a waterborne pathogen was the culprit,[48] a scenario that sounded likely, given that the mass bleaching was occurring at about the same time as the bacteria-caused die-off of *Diadema*—and most noticeably in the same general region where that other epidemic began. But as the bleaching episode continued, it became apparent that something else was at work. The sea urchins were dying along a pathway of Caribbean currents. Coral bleaching sometimes followed that pattern, but frequently the spread appeared haphazard, striking reefs not linked by currents. Scientists reported bleaching on reefs as far apart as French Polynesia, the Florida Keys, southern Japan, and Indonesia.[49] And it wasn't a single species of coral that was affected, as was the case in the *Diadema* die-off. In fact, the bleaching wasn't limited to corals at all. Other creatures living in tandem with zoox also bleached, including sea anemones, sponges, soft corals, and even mollusks.[50]

Many of the corals that didn't bleach died anyway, as a result of the "downstream" effects of the phenomenon. Off Uva Island, for example, large stands of unbleached massive corals were attacked and destroyed by an invasion of the voracious crown-of-thorns sea star. Normally the sea stars were kept away from those corals by small but aggressive crabs and shrimp. The guardian crustaceans live symbiotically with another group of corals, the *Pocillopora*, a genus of branching corals that normally provide the crustaceans protection from predators, as well as nutrients from coral-produced mucus.[51] Glynn noted that an average of twenty-two crustaceans lived in a typical colony of healthy pocilloporid corals.[52] At Uva Island, the massive corals lived within a protective ring of pocilloporids. When the outer corals first bleached, their mucus production dropped off and many of their crustaceans either left or died. Glynn found half the number of guardian crustaceans in bleached colonies of pocilloporids, and generally none in colonies that had died as a result of bleaching.[53] When entire tracts of the pocilloporids died (mortality rates reached 100 percent in some areas), their crustaceans died, too, leaving the way open to an attack by crown-of-thorns sea stars.[54] In 1991, seven years after the attack began, Glynn returned to Uva Island and found that the sea stars were continuing to devour giant coral heads that were a century or two old.[55]

Even as scientists continued to monitor the path of destruction caused by coral bleaching, they were struggling to determine what was causing the bleaching. Pooling data and reviewing literature, biologists came up with a large number of possibilities, all grouped together under the general heading of "stress." Corals lost their zoox when seawater salinity was too high or too low, when they were exposed to toxic substances or high levels of sediments, or when they received too much or too little light—to list just a few of the more common causes. In 1983, Glynn was probably the marine biologist most qualified to speculate on the bleaching's cause. He ruled out many factors as unlikely, but was unable to say with any certainty what was inducing the bleaching. He did, however, conclude that "it seems possible that they may be related to the strong El Niño event."[56]

There was good reason to suspect the elevated seawater temperatures caused by the ENSO event. Many if not most coral species have adapted to life within a surprisingly narrow range of conditions: salinity, nutrients, sediments, temperature. Even a slight fluctuation in any of these factors can mean the difference between life and death. Corals are, in a sense, the high-wire artists of the sea, where only a slight misstep spells doom.

In the laboratory, scientists found that corals bleached after just two days when the water temperature was raised five to seven degrees Fahrenheit above their normal seasonal highs. But corals were even more sensitive to slight temperature fluctuations over longer periods. An increase of just two degrees, sustained for several weeks, could bleach and kill corals.[57]

After gathering and analyzing the data for a year, Glynn was ready to go farther out on the limb, linking the mass bleaching with the most recent ENSO event. His article "Widespread Coral Mortality and the 1982–83 El Niño Warming Event" was seminal for coral scientists, even though Glynn couched his conclusion in the cautious language of science. He wrote that the mass bleaching and mortality of corals around the world and the ENSO warming "may not be entirely independent events."[58]

If the story ended here, it would be the tale of a catastrophe, for at least some of these bleached reefs will, in all likelihood, never return. But the 1982–83 ENSO event was the beginning of the story, not the end.

As time went on, it became clear that many pieces of the bleaching puzzle fit the ENSO pattern, while in other cases the connections were more tenuous. For example, bleaching had already begun on the Great Barrier Reef *before* the 1982–83 ENSO event.[59] Then, in 1986–87, another, weaker, ENSO event took place.[60] Oddly, a study found that bleaching in 1987 was "the most severe and widespread ever observed,"[61] affecting reefs in Australia, throughout the Pacific and Indian Oceans, and all across the Caribbean. And even long after the ENSO event had subsided, bleaching spread throughout the Caribbean.[62]

In 1987, with concern mounting about these events, the U.S. Senate held a special hearing on bleaching.[63] "The mass bleaching of Atlantic corals is frightening," NOAA scientist Robert Wicklund told the senators, "particularly considering its extent and range."[64]

When bleaching continued in 1990, a non-ENSO year, the Senate invited Dr. Wicklund back to testify.

"I was concerned with the extent of the bleaching event in 1987," a somber Wicklund reported, "but today I am dismayed. Our coral reefs are in peril and are disappearing at an alarming rate."[65] In fact, Wicklund continued, bleaching on some reefs in 1990 surpassed 1987 levels—even without an ENSO event.

The common element still appeared to be increased water temperature; there was little disagreement over that. Far more contentious, in part because the implications were so far-reaching, was the question of what was causing elevated temperatures, if not ENSO events.

Many scientists began to suspect human-induced global warming. While there was much debate during the early-to-mid-1980s over whether or not the planet was experiencing a warming trend, by the end of the decade there was near unanimity among scientists on the question: global warming has been occurring for the last hundred years, largely, if not entirely, as a result of human activity.

The warming trend began in the nineteenth century, as humans first began to burn large amounts of fossil fuels, primarily coal and oil. When those fuels are burned, carbon is released into the atmosphere in the form of carbon dioxide. Scientists measuring the amount of carbon dioxide in the atmosphere found that levels of the gas there had increased by nearly one-third over the past century.[66] Like the windows of a greenhouse, the layer of carbon dioxide allows solar energy in, but doesn't allow the heat that is generated to escape (which is why carbon dioxide is often referred to as a "greenhouse gas"). As a result, Earth's mean temperature has risen between one and two degrees Fahrenheit. That may not sound like a lot, but the effects are potentially catastrophic, particulary for coral reefs, which already exist within a narrow temperature range. Because of the complex feedback loop between ocean and atmosphere, greenhouse gases could also be responsible for more frequent and more severe ENSO events.

Was global warming really responsible for global bleaching? A consensus that this was indeed the case began to grow. In April 1990, Lucy Bunkley-Williams and Ernest Williams Jr. made the link in *Natural History* magazine.[67]

Before the Senate hearing in October 1990, Ernest Williams Jr. testified, "The first proof of global warming may well come from the bleaching of the fragile and highly sensitive coral reef system. . . . These disturbances may actually be alarms from the drastically changing and deteriorating marine environment."[68]

When asked by senators to comment on this connection, Thomas Goreau, president of the Global Coral Reef Alliance, an organization of scientists, said that while there was not enough evidence to say with scientific certainty that the so-called greenhouse effect was to blame for the recent, and continuing, wave of global coral bleaching, "it certainly does look like that to us," adding that "personally, I do not doubt there is a connection."[69]

Bleaching continued in 1991. On June 12, 1991, an event occurred that would lend credence to the link between global warming and coral bleaching. On that day, Mount Pinatubo in the Philippines ended six hundred years of dormancy, erupting in a spectacular explosion and blasting seven cubic kilometers of rock and ash into the air and sending gases some twenty-five miles high.[70] Unlike greenhouse gases, these aerosols reflected sunlight back into space, lowering temperatures around the globe. If mass bleaching resumed after the effects of Mount Pinatubo wore off, some scientists suggested, global warming would appear to be the likely culprit.[71]

And that's what happened.

In 1995, worldwide temperatures began to climb again, with ocean surface temperatures reaching their highest levels since 1984.[72] Coral reefs around the world began to bleach, including many that hadn't suffered during previous episodes.[73] One of the most alarming developments came in September 1995, when the Caribbean's longest barrier reef, off the coast of Belize, experienced mass bleaching for the first time.[74]

By mid-1996, Thomas Goreau spoke for many scientists when he stated, "There's no doubt that global warming causes bleaching."[75]

Evidence of this tragedy grows each year. Mass bleaching events were once again reported around the world in late 1996, from Bahrain in the Persian Gulf[76] to the Solomon Islands in the

West Pacific.[77] There now seems little doubt that what was once a cyclical but relatively rare event has become a near-annual occurrence—and may well be for the foreseeable future. Even as I write this, in the late fall of 1997, the Climate Prediction Center at NOAA is issuing bulletins warning of a new ENSO event.[78] Sea surface temperatures, they report, are at their highest levels in fifty years.

Bill McKibben wrote an eloquent and impassioned book on the implications of the greenhouse effect in 1989, back when there was far less scientific evidence than there is today that global warming is a human-induced phenomenon. His thesis, that human industrial activity, primarily the burning of massive quantities of fossil fuel, has caused a break with our age-old relationship with nature, was suggested in his provocative title, *The End of Nature*. Near the end of that book, McKibben tries to imagine an alternative to the apocalyptic course on which we seem to be embarked. He poses the following question:

"What would it mean to our ways of life, our demographics, our economics, our output of carbon dioxide and methane if we began to truly and viscerally think of ourselves as just one species among many?"[79]

Many readers saw McKibben's question as heretical, not to mention humbling, with implications that were simply unacceptable. But others saw in it a liberating and even exhilarating perspective. Either way, whether we interpret his question as optimistic or pessimistic is beside the point. For his premise is true. We really are "just one species among many"—and sooner or later we must act the part. To paraphrase Gabriel García Márquez: A species that lives in solitude gets no second chance on Earth.

15

Unweaving the Outer Braids

Man's fingerprint is found everywhere in the oceans.

—*The State of the Marine Environment*, United Nations Environment Programme, 1990

It is well after midnight when my flight finally touches down at Ngurah Rai International Airport, on a narrow spit of land at the extreme southern edge of Bali. The driver from my hotel—a polite young man, clean-shaven even at this hour, with a stylish haircut and a businesslike demeanor—stands outside the terminal, patiently smoking a fragrant Kretek in the humid tropical air. In his free hand he holds a tattered piece of paper with a rough but recognizable version of my name, something like DOSA BAVDISSON. I toss my bags into the back of the car, climb in after them, and collapse. Thirteen hours earlier I had boarded a flight in Townsville, Australia, flown to Cairns on the northern tip of the continent, and caught a flight west over a thousand miles to Darwin, where I'd boarded my jet bound for Bali, another thousand miles northwest over the Indian Ocean.

"Okay I put on music?" asks the driver.

Sure, I tell him, and he puts in a tape. I silently pray that this isn't one of the pirated tapes of bad American pop music, ubiquitous throughout Southeast Asia. The Balinese gods must be feeling generous toward foreigners tonight. Instead of a poor-quality dub of Madonna singing *Like a Virgin*, the car suddenly fills with the raucous and mystical music of a gamelan orchestra, the traditional music of the Indonesian archipelago. The Balinese style is played louder, faster, and with more dissonance than anywhere else. Cymbals crash, drums throb, gongs gong, and through it all are the spine-jarring jabs from the *gambang*, an onomatopoeically named instrument with a bronze base and keys of bamboo or wood. The *gambang* has been likened to the xylophone, which is like saying a hungry tiger is similar to a housecat. Listening to the Balinese gamelan is like being caught in a rip current or tossed about by a huge breaking wave. There is nothing for the western mind to grab hold of. One guidebook warns that some people find gamelan music "curiously melancholy, even disturbing."[1] Your best bet is to let go and just enjoy the ride.

Revived by the music, I roll down my window and watch the Balinese countryside rush by. It is the first night of the new moon, an auspicious time for travel, perhaps, but not for viewing the countryside. At first I can make out only objects close to the road. A few buildings. Tangles of mangroves. As my eyes grow accustomed to the dark, I see more. Water buffalo stand in the warm night, swiveling their enormous heads as the blaring gamelan music sweeps past. Behind these great beasts, and beyond the mangroves, the heaving waters of the Indian Ocean glisten with reflected starlight. Everything we pass is silhouetted against the sea: dark shapes cast upon a light background. It is impossible to watch this parade of silhouettes without thinking of shadow puppets—another traditional Indonesian art. With the gamelan music crashing down around us, we pass shadow puppets of more water buffalo, shadow puppets of civic statues, shadow puppets of mangroves. Then there is a break in the mangroves as we pass a small rectangular field, its surface glinting with water. More of them. Soon the mangroves are gone entirely.

"What are the fields?" I ask the driver. "Rice?"

He turns his head and surveys the landscape.

"No rice," he yells over the music. "*Udang*."

"What are *Udang?*"

"*Udang?*" he repeats, and thinks for awhile.

He holds up one hand, bringing the fingers close together and making a cup of them. Then he flutters his fingers in a syncopated crawling motion.

"*Udang! Udang!*" he shouts.

"*Udang?*" I repeat, as if saying the word again will render it understandable.

The driver is quiet for a while. You can tell the exact moment that the English word leaps into his head. He suddenly grips the wheel more tightly and jerks his head toward me.

"Shrimp!" he calls out, a triumphant smile illuminated by the blue glow of the dashboard light. "*Udang*, shrimp!"

He makes his hand-sign again. This time it's obvious: a shadow-puppet shrimp. His fingers are the legs, pulsating in the water.

"*Udang!*" he says again, smiling. The shadow-puppet shrimp becomes a pointer that directs my gaze back out of the window into the glimmering night. "Lots shrimp grow there!"

That is something of an understatement. Nearly everywhere along the coastline of Southeast Asia, shrimp are raised in huge numbers. It has been called the greatest economic success story of the last several decades, as shrimp ponds are created to satisfy the voracious appetite for the delectable crustaceans in the United States, Europe, and Japan. Less than two decades ago, shrimp were a luxury food. Now you can find them everywhere: heaped in gleaming piles on ice in the fish section of many supermarkets; sold frozen by the bagful in smaller grocery stores. Even the McDonald's at your local strip mall regularly features a shrimp concoction. And, of course, it's served in nearly every restaurant in America, from medium-priced to upscale, in all sorts of ways: blackened Cajun style, sautéed in garlic sauce and served over angel-hair pasta, brushed with butter and broiled on a grill. In the past decade alone, American shrimp consumption has doubled—to nearly a billion pounds a year, about two and a half pounds for every American.[2] Half of this enormous total comes from shrimp raised in ponds in Asia and Latin America.[3]

The basic practice is an ancient one in Southeast Asia, going back many hundreds of years. Shrimp and fish were raised for local consumption in rice paddies during nongrowing seasons. The juvenile shrimp that live in the brackish waters of the coastal mangrove forests would simply pour in when the fields were allowed to flood with this water. The shrimp and other crustaceans grew fat on the plankton that followed them in. When planting time came, the shrimp were harvested. Then the brackish water was replaced with fresh, and a new crop of rice was planted. The operation was always small-scale—simply a way to add a bit of extra food or income during the nongrowing season.

That all changed in the late 1970s and 1980s.

"The Indonesian government decreed, 'Thou shalt make shrimp ponds,'" says coral scientist Evan Edinger, who has tracked the growth of these ponds, called *tambaks* in Indonesia. "First the government said it was to increase protein in rural diet. Then they said it was to increase export income."[4]

Whatever the real reasons, almost overnight and all across Southeast Asia, shrimp cultivation became a huge business, fueled by international lending agencies, government aid, corporate wealth, and the desire by small farmers to increase their wealth, and by wealthy landowners to grow richer still. Many shrimp farms are the property of absentee landowners, who, in central Java, are known as "farmers with ties." Worldwide, farm-raised shrimp production jumped from 92,000 tons in 1982 to 808,000 tons in 1994,[5] an eightfold increase in a little more than a decade. The total dollar value of the farm-raised shrimp industry is now conservatively estimated at about $8 billion annually.[6]

But these impressive gains weren't made possible by just a few traditional *tambaks*. More *tambaks* were needed, many more. And so, many more were built. But it still wasn't enough. These mom-and-pop-type arrangements weren't productive enough to meet the international demand for shrimp. Besides, if you can earn a few extra rupiah with a regular *tambak*, think what you could make with a modern, factory-style *tambak*. Enter the super-*tambaks*—basically shrimp ponds on steroids. These ponds are stocked with up to a hundred times the density of shrimp found in traditional *tambaks*.[7] Of course, there's not enough food naturally occurring in the ponds for that many shrimp, so the new "farmers" add fertilizer to the water to stimulate algae growth. Many supplement the shrimps' diet with commercial feed. And since a pond jammed with tens of thousands of shrimp is a perfect breeding ground for disease, gallons of antibiotics are poured into these intensive shrimp ponds. Even with these medicines, shrimp raised in super-ponds often fall prey to disease, with hundreds of contiguous ponds wiped out in a short period. Taiwan, once a leader in shrimp production, saw virtually its entire industry destroyed when a disease raced through its ponds between 1987 and 1989.[8]

Infected ponds must be abandoned. But not to worry: there are always more mangroves nearby. And so the owners simply move down the coastline a bit, clear another section of mangrove forest, and begin the process again.

What makes these superponds such an environmental nightmare is the fact that even without a ravaging disease, they become so filled with chemical pollutants and nutrients from decomposed feed and shrimp fecal matter that they must be abandoned after only five to ten years.[9] On the island of Java, where most Indonesian superponds have been set up, an estimated 70 percent have already been abandoned.[10] These former superponds are on their way to becoming the most common toxic-waste sites of the developing world. Nothing can grow in the polluted soil, certainly not more shrimp or fish, or rice or other food crops. It is even difficult to get mangroves to regrow in this soil.

"*Tambak*s are really bad news," says Evan Edinger, the coral scientist who teaches at a university in the central Javanese city of Semarang. "In my classes, when I show photos of *tambak*s, I tell students, 'Here's public enemy number one.'"

The practice has been called "rape-and-run" aquaculture,[11] the coastal-zone equivalent of "slash-and-burn" agriculture, and the resource that has allowed it to continue for two decades is rapidly being depleted. The supply of mangrove forests only seemed endless. Today, thanks in large part to shrimp-pond conversion, the end of this underappreciated and little-understood ecosystem is in sight. If this important outer braid of the tropical coastal system is to be saved, major changes are going to have to occur, and soon.

"I'm afraid that the ecological function of mangroves is being rapidly eliminated by humans," says Anthony Viner, a New Zealand native who is team leader of the Mangrove Rehabilitation and Management Project in Sulawesi, Indonesia.[12]

In 1965 over a million acres of mangrove forests lined the shores of the Philippines.[13] Today, three-quarters of them have been destroyed.[14] In Thailand, mangroves suffered a similar decline, dropping by more than 50 percent in just three decades,[15] with the bulk being lost in the last decade alone.[16] Indonesia, blessed with some of the largest tracts of mangrove forests in the world, has already lost a third of its original forests, mostly to *tambaks*.[17] Fully half of all mangrove lands in Southeast Asia have been destroyed over the past fifty years[18]—again, primarily because of the construction of shrimp ponds[19] (many which have already been abandoned).

The widespread destruction of mangrove forests is creating tremendous hardship for the people who live in coastal communities and who have depended on mangroves for several hundred years. Given the myriad benefits that mangroves hold for local residents, it's ironic that many outsiders still view mangroves as "wastelands." Wood from mangrove forests is the primary fuel source in many areas. When a new plow is needed, it can be carved from a mangrove tree. Several fruits and medicinal plants are harvested from these forests. As ICLARM's John McManus points out, "With ten hectares of mangroves, about a thousand people benefit. With ten hectares of shrimp ponds, maybe one or two people will benefit."[20] Since most mangrove forests are publicly owned, local people have traditionally had access to them. But once developers buy the land from the government, the mangrove resources are forever lost to the community.

Because mangroves are a transitional zone linking land and sea, their marine benefits are equally important. In many areas, people depend on mangroves for the enormous quantities of mussels, cockles, and crabs that thrive among their roots. Mangroves are also important nurseries for fish and crustaceans harvested by local people. When shrimp ponds replace mangroves, those nurseries are destroyed, resulting in drastically reduced catches for local fishermen. Then there is the added problem of contamination of local fisheries by the polluted waters of superponds. Following the conversion of mangroves to shrimp ponds in a section of Bangladesh, fishermen reported a drop of 80 percent in their average catch in waters surrounding the new ponds.[21]

Mangroves also provide coastal people with protection from storms and tidal waves, an insulating effect that is lost when the land is turned into shrimp ponds. In 1991, for example, a tidal wave killed thousands of villagers in a section of India that had recently been converted from mangroves to shrimp ponds. The same region hadn't suffered a single death when a similar tidal wave struck in 1960—when mangrove forests still served as a buffer zone.[22]

Less understood, but possibly even more important in the long run, are the downstream ecological effects of the destruction of this outer braid of the coral reef. What is known is that mangroves prevent land-based pollutants and sediments carried by rivers from flowing out to the coral reefs. When this natural filter is removed, waterborne particles flow unimpeded out onto reefs, promoting algal growth at the expense of corals or simply smothering the living corals beneath a layer of sediment. And, of course, shrimp ponds add their own pollutants to coastal waters, in the form of nutrients and chemicals that almost certainly damage reefs.

Out past the mangrove forests (or, all too often, where mangroves used to grow), another of the outer braids, sea-grass beds, are also facing global threats. This is especially true in the "magic triangle" of biodiversity where coral reefs reach their apogee. Miguel Fortes, a professor at the University of the Philippines and a leading authority on sea grasses of Southeast Asia, estimates that the Philippines have lost as much as half of their sea-grass meadows in the last fifty years.[23] Indonesia has fared almost as badly, with between 30 percent and 40 percent destruction, and around the heavily populated island of Java, probably 60 percent of sea-grass beds have been destroyed.

As with the destruction of mangroves, the widespread devastation of sea-grass beds has a number of serious implications. Sea grasses support a rich variety of other life, ranging from microscopic algae to huge marine mammals. More than a hundred varieties of epiphytes (plants that grow on other plants) have been found on a single sea-grass genus.[24] The sea grasses, together with the epiphytes on them, produce a tremendous quantity of nutrients, the food supply for the many mollusks, sea cucumbers, and sea urchins that are in turn harvested and eaten or sold by coastal people.[25] The dugong, an endangered marine mammal closely related to the manatee, requires enormous amounts of sea grass for its diet. (A captive dugong in the Surabaya Zoo in central Java eats sixty-six pounds of sea grass each day.[26]) The green sea turtle also requires large quantities of sea grass. And certain varieties of seahorses—exquisite and bizarre creatures that are being rapidly hunted out of existence—depend on sea grasses, not for food but for habitat.[27]

One of the most important ecological roles of sea-grass beds may be as nurseries for many varieties of coral reef fishes. Juveniles that would be vulnerable to predation on the reef spend a safe period in sea-grass beds. According to one estimate, 70 percent of commercial fish spend at least some part of their life cycle in sea grass.[28] Over three hundred species of fishes, mostly juveniles, were recently found living in just three Indonesian sea-grass beds.[29] The destruction of this important habitat could have important implications not just for commercial fisheries, but for the ecology of coral reefs, as the food web is disrupted.

Extensive sea-grass beds also act as a buffer, similar to mangroves, filtering out harmful sediments before they have a chance to reach the reef.[30] With mangrove forests destroyed, sea-grass beds are even more critical in protecting reefs from land-based sediments. The problem, of course, is that sea grasses are most subject to destruction in the same areas in which mangroves are eliminated—that is, near human population centers. An estimated half of all sea-grass beds within thirty miles of urban coastal areas are now gone.[31]

As good as they are at settling sediments, even sea grasses can handle only a certain density of particles in the water before they, too, suffer. Mining, logging, and coastal construction can cloud waters over sea-grass beds, cutting out light needed for photosynthesis and laying waste to large tracts of sea grasses. The polluted waters from shrimp superponds have also been responsible for major sea-grass die-offs, and shallow beds have been wiped out by a variety of destructive fishing practices, including blast and cyanide fishing and trawlers that drag large nets across them.[32]

The fact that many sea-grass species are slow to recover from physical damage makes their destruction all the more serious. Tony Larkum, the scientist who first opened my eyes to the marvels of sea grasses as we snorkeled together in the San Blas Islands off Panama, makes this point when we next meet, four months later and some seven thousand miles away, on a stroll down one of the many tree-lined footpaths in Kurnell National Park, just south of Sydney, Australia.

"This is where it all began," Larkum says, as we stop to admire a marker honoring the spot where Captain James Cook anchored the *Endeavour* in April 1770, and for the first time set foot upon the Australian continent. "It's your typical history of exploitation and neglect," Larkum says, squinting into the morning sun as he gazes out over the Botany Bay. His comment isn't meant as the sweeping indictment of Euro-Australian history that it sounds like (although there are those who would argue it applies). Larkum is remarking on the destruction of sea-grass beds within Botany Bay. Two hundred years ago, lush meadows of a sea grass called *Posidonia australis* covered nearly the entire bottom of the 840-acre embayment. Then a slow decline of the sea grass began. In the nineteenth century, boats dredging for mud oysters tore up sections.[33] Next, industrial pollutants and sewage from the growing city of Sydney (six miles to the north) were dumped into the bay, killing more *Posidonia*.[34]

It wasn't until relatively recently, however, that the destructive process shifted into high gear.

"We have aerial photographs of Botany Bay since the 1940s," Larkum says. "They show a slow decline until 1975, when the port was put in. And then a precipitous decline of *Posidonia*." The dredging for Port Botany allowed large waves to bear down on sea-grass beds, destroying large areas. Today, for all practical purposes, *Posidonia* has been wiped out in the northern section of Botany Bay. Larkum characterized the condition of sea grass in the southern part of the bay in a journal article: "By 1987, the once continuous meadows of *Posidonia* consisted of a number of fragmented beds."[35]

All of this is particularly ironic in an area that was named Botany Bay by Captain Cook's naturalist because of the extreme diversity of the plant life he found here two hundred years ago. "It must have seemed like Eden to them," Larkum observes. "And now you have less sea grass and more areas of sandy bottom."

What particularly upsets the generally unflappable Larkum is that so little thought was given to the consequence of actions, such as dredging, on the sea grasses—even though Botany Bay is a major nursery for the New South Wales fishery.

"It's a rather commonsense argument," he points out. "If you're disturbing an environment, you damn well should put a little money into understanding how it works."

That evening, back at his house in a quiet residential section of Sydney, in the shadow of the university where he's taught for decades, Larkum sets up his slide projector so he can show me a series of bizarre images: aerial views of *Posidonia australis* beds down in Jervis Bay, a hundred miles south of Sydney. What's strange about the photos is that the beds of pure green are punctuated by a series of enormous perfect circles of yellowish sand.

"The circles first showed up in these photographs taken around 1967," Larkum narrates. No one knows how the circles got there—or at least the Australian public doesn't know. A nuclear reactor for that area was on the drawing boards back in those years, and the most popular theory explaining the gaps in the sea-grass beds is that they are the result of explosions set as part of seismic tests to determine the feasibility of locating a reactor there.

"But however they got there," Larkum stresses, "the important part is that they're still there, some thirty years later." He switches to the next slide, which he says was taken recently. I'll have to take his word for that, because the image appears identical to the first one. According to Larkum, that's because the sea grass has regrown into the circles at a rate of only about four inches a year.[36]

The destruction of sea-grass beds—whether caused by explosions, dredging, or shrimp ponds—can take many decades, or more, to remedy. And in that time, what other biomes are damaged and perhaps lost? There are no good answers to this question.

Sea-grass meadows have been called "the least studied among the living resources of Southeast Asia." But scientists trying to rectify the situation find themselves racing against the clock. As Miguel Fortes concludes in his report on this precious ecosystem, "The current rate that we are acquiring information on sea grasses is less than the rate of loss of these resources."[37]

❄

I have a chance to consider all these problems on my way back to the airport to leave Bali for the island of Sulawesi. I pass the labyrinth of *tambak*s once again, this time in full daylight. My driver, who is named Nyoman, is just as young and even more talkative than the man who first picked me up at the airport. He notices that I am staring at the hundreds of ponds lining the road.

"*Tambak*s," he says. "They raise many prawns there."

"Yes," I say, "I know."

We fall into an easy conversation.

"You married?" he asks. I tell him that I am, and show him pictures of my wife and my children. He points to the picture of my two-year-old son, whose tangle of blond curls makes his sex ambiguous.

"*Anak laki-laki,*" I say, Bahasa Indonesian for "son."

"You lucky," he says, a truth I readily acknowledge.

"I get married in few months," Nyoman says proudly, sitting up a little taller in his seat. He tells me that his family and his future wife's are farmers in central Bali.

"Do they raise prawns?" I ask.

Nyoman laughs uproariously at the idea. "No, no, no," he says. "My family is poor. For *tambak*s you must be rich. Like you."

As a freelance writer back in America, I'd hardly call myself rich. But by world standards there is no doubt that I am. My wife and I own our own cars, and we are buying an old but comfortable house, with running hot and cold water and two bathrooms. I travel halfway around the world to research a book (although after I arrive home, I'll be greeted with a credit-card bill that will take me months to pay off). Embarrassed by the truth of Nyoman's statement, I look back out the window. Just then we pass a small, muddy field studded with scrawny mangrove saplings, a token Japanese experiment in *tambak* reclamation. Wetlands are hot items with international lending agencies right now, and so they're fighting each other to buy a few showcase wetland projects. This is a perfect example of the result: a tiny field of pathetic-looking mangroves wilting in the sun, surrounded by thousands of super-*tambak*s. In the middle of the sorry field there is a sign rising out of the mud. In bright red letters it proclaims, "WE ♥ MANGROVES!"

"What is funny?" Nyoman asks. It's only when he asks this question that I realize I laughed out loud.

I doubt that my little Berlitz phrase book has the Bahasa Indonesian word for "irony," and Nyoman probably doesn't know the word in English, and so in the end I have to settle for saying merely, "Nothing."

Which, of course, is the truth.

16

Once More to the Keys

It is grotesque to see tourists photographing and marveling over small surviving coral remnants of once great reefs drowning in sediment as if they were little jewels. One would hardly photograph terminal patients in a cancer ward as glories of the human condition. But people do not know any better about coral reefs, so there is no public perception of the magnitude of our loss.

—Jeremy Jackson, Smithsonian Tropical Research Institute

Everyone's taken the reef for granted. But there's only so long that you can do that.

—DeeVon Quirolo, Reef Relief

Everybody loves a good shipwreck.

The pioneering residents of Key West surely did. It's what turned their sleepy island village into one of the richest towns per capita in the United States in the early nineteenth century.[1]

"A wreck was the most wished for and thoroughly enjoyed thing that could happen," recalled Mary Munro in 1880.[2]

The merchant ships that foundered by the hundreds on the nearby reefs were floating treasure troves, their holds filled with fine linens, crystal glasses, silverware, and bottles of wines. Under U.S. law, the wrecks and everything on them belonged to whoever reached them first, and so the competition to claim the broken ships was fierce, with as many as thirty salvaging boats racing each other out to the reef, often in the same foul weather that caused the crackup in the first place.[3] The wreckers auctioned off most of the goods, adding more than $1.5 million annually to the small island's economy.[4] In one year alone, seven hundred vessels were dashed upon the reefs.[5] Life was good. Wreckers kept some of what they salvaged, and many of the mansions built during those early years were made from ships' timbers and furnished with the goods hauled from the damaged vessels. And if business was slow, particularly in the days before law enforcement was present in the Keys, wreckers sometimes lit false beacons to guide ships to their doom.[6] The number of ship groundings dropped dramatically when the government built a series of lighthouses along the reefs, and sent military authorities to crack down on the wreckers. Key West had to turn to other enterprises, such as cigarmaking, sponge-collecting, and turtle-catching, but ships still continued to run aground on the reefs from time to time.

The media also love a good shipwreck, particularly if it occurs on a living reef.

A major wreck has all the elements of a great news story. The tragedy is compelling and easily conveyed: SHIP HITS CORAL REEF! The damage is obvious and dramatic. Run a four-hundred-foot freighter aground on a reef of delicate corals (as happened in August 1984 on Molasses Reef, off Key Largo), and you end up with "a graded roadbed covered with a veneer of coralline debris."[7] And, of course, there are the great visuals, a critical element in an age that demands its disasters televised—the obligatory helicopter shot of the vessel as it lists on its crumpled hull, smaller boats churning about like water beetles scurrying around a dead fish.

It may be going a bit far to say that the author of an Interior Department report on potential oil spills on the Keys reef tract loved a good shipwreck, but he or she did manage to find the sunny side of a potential maritime disaster in the Keys, citing "beneficial effects resulting from expenditures made by research and media personnel and curious onlookers."[8]

Even those who love the reef are able to turn a shipwreck to their advantage, by focusing public attention and galvanizing political will on threats to the 220-mile-long Florida reef tract. In fact, the Florida Keys National Marine Sanctuary, the second-largest coral reef protected area in the world (after Australia's Great Barrier Reef Marine Park) owes its existence in large part to a series of three boat groundings that occurred within one eighteen-day period in 1989.

The first occurred at around two-thirty in the afternoon on October 25, 1989, when a southbound oil-supply ship, the 155-foot *Alec Owen Maitland*, ran aground in less than ten feet of water in the Key Largo National Marine Sanctuary, toward the northern end of the Keys reef tract. The ship scraped bottom for three hundred feet, toppling some coral colonies and pulverizing others. After grinding to a stop, the crew attempted to free the *Maitland* by turning it. Bad idea. This resulted in another two hundred feet of devastation. Finally the boat became firmly wedged in less than a yard of water, creating another two-hundred-foot area of destruction.[9] It was later determined that the *Maitland*'s captain wasn't in control of the vessel at the time of this incident. He had illegally turned the helm over to the ship's first mate and his brother, the ship's engineer. Both of them were drunk at the time of the grounding.[10]

Just five days later, a Yugoslavian cargo vessel, the *Mavro Vetranic*, caused three acres of damage when it, too, ran aground. This time the grounding occurred at the southern end of the reef tract, just off Fort Jefferson National Monument Park.

On November 11, the Greek freighter *Elpis* was carrying a load of sugar from Holland to Tampico, Mexico. Shortly before midnight, while sailing under a nearly full moon and within a quarter-mile of a reef marking light, the *Elpis* ran aground on The Elbow reef, again within the Key Largo National Marine Sanctuary. Ninety-five percent of corals struck by the 470-foot boat were killed. As in the case of the *Alec Owen Maitland*, much of the damage was done in attempts by the *Elpis* to "power off" the reef by using its own engines. The ship's propeller created twin blowholes, one as it attempted to move forward, the other as it tried to back up, all to no avail. The *Elpis* finally came to rest in a section of the reef dominated by soft corals with many small hard coral colonies. Perhaps the most destructive element of the grounding wasn't the damage to living corals, but to the reef framework itself, which was reduced to rubble in some places.[11]

The timing of these groundings couldn't have been better. Images of oil-covered birds and sea otters from the *Exxon Valdez* grounding in March 1989 were still fresh in the public's mind. The giant oil tanker had run aground in Alaska's Prince William Sound, sending 240,000 barrels of oil cascading into the pristine environment, causing one of the nation's worst ecological catastrophes. Coming just seven months later, the three Florida groundings were linked in the public's perception to the *Valdez* tragedy.

Conservationists and others concerned about the Florida reef tract had been fighting battles to protect the reef for years, most recently to prevent oil exploration and drilling in the Keys. The groundings suddenly placed the complex issues of reef preservation on the national agenda, albeit in an oversimplified fashion. The Florida reef tract had been given its fifteen minutes of fame. Hearings were immediately planned for the creation of a Florida Keys National Marine Sanctuary, introduced in the House of Representatives by Florida congressman Dante Fascell. When those hearings were opened in May 1990, speaker after speaker mentioned the three boat groundings. It began with the opening statement of the chairman of the subcommittee on Oceanography and the Great Lakes.

"Last fall," intoned Chairman Dennis Hertel, "three separate tanker groundings were reported along the coral reef of Florida within a three-week period. These incidents brought to light the need for action to be taken to protect the Keys coral reef and the fragile ecosystem which it sustains."[12] (Actually, none of the vessels in question was a tanker, but the *Valdez* was, and perhaps it's not too cynical to suggest that Chairman Hertel used the word purposely to recall that earlier disaster.)

The very first witness before the subcommittee was Florida senator Bob Graham, sponsor of the bill in the Senate. Graham was straightforward about the connection between the groundings and the legislation: "Congressman Fascell's bill was offered in response to a highly publicized grounding of large vessels. . . ."[13]

Other witnesses spoke about potential horrors of shipwrecks on the reef, and in the end there is little doubt that it was that threat which prompted Congress to pass Public Law 101-605 on November 16, 1990, creating the Florida Keys National Marine Sanctuary. Several environmental threats are mentioned in the act, but the very first one is "vessel groundings."[14]

❄

Nearly six years later, Dr. Brian Lapointe sits on a stool at Herbie's, a funky eatery on Marathon Key, wolfing down a fish sandwich and snorting at the threat posed by boat groundings.

"Hmppphh," he growls through a mouthful of mahi-mahi, when I ask about the damage done by shipwrecks. He swallows and says, with the brash self-confidence that is Lapointe's trademark, "They were dealing with a trivial pursuit. It was bullshit."

Lapointe, a marine biologist with the Harbor Branch Oceanographic Institute in Fort Pierce, Florida, isn't alone in this view, though few are willing to speak as bluntly as he does.

During the 1989 House hearing, John Ogden, director of the Florida Institution of Oceanography in St. Petersburg, Florida, pointed out that "collisions between ships and coral reefs are dramatic but relatively insignificant to a reef over two hundred miles long."[15] And Craig Quirolo, head of the Key West–based grassroots group Reef Relief, also downplayed the importance of boat groundings in his testimony before the subcommittee.[16]

Most scientists and activists working on the Florida reef tract agree with Lapointe on this point: boat groundings are not the *real* problem. Where they differ—and the disagreements can turn nasty—is over the question of what the real problem *is*. Overfishing? Global warming? Physical damage by divers? Perhaps, some point out, what's happening in the Keys is nothing more than the downside of the normal rise-and-fall cycle that corals have been experiencing for millennia. And that's what makes the reef tract here so important (aside from the fact that it's the only barrier reef in the continental waters of the United States): the bewildering number of questions about what's causing them to decline. If answers are found here, they may have relevance for reefs around the globe.

"This is one of the hot spots," says coral physiologist Erich Mueller, who lives and works in the Keys. "It's the front line in the battle to save the environment."[17]

There is little doubt that the Florida reef tract is sick. If it were human, the patient would be in the intensive care unit. Nearly every year a new disease is discovered on the corals: Yellow Band disease; Black Band disease; White Plague, type 1; White Plague, type 2; Red Band. A whole series of opportunistic diseases are attacking corals. James Porter, a zoologist at the University of Georgia, was the first to document widespread coral decline in the Keys over a period of years. Porter monitored six different reef sites from near Miami all the way down to Looe Key (twenty-five miles east of Key West) between 1984 and 1991. What he found was alarming: five of the six sites had lost hard coral cover, at an average rate of around 5 percent per year.[18] At that rate, the Florida reefs had only a couple of decades left.

Some scientists have questioned whether Porter's findings are applicable to the entire reef tract. Others have raised the possibility that what Porter documented was a short-term anomaly. Still, his work raises serious questions about the health of the reef. Porter has a theory about what is causing the deterioration: a decline in water quality. At least in its broad outline, this is an echo of what Brian Lapointe had been saying—*hollering* is probably the better word for it—for years. Water quality is *the* issue, insists Lapointe. To understand what's happening to the reefs here, you have to broaden your view to include a whole series of distant braids.

Since Louis Agassiz first conducted a scientific survey of these waters back in 1851, the effect of the Florida Current, the huge body of warm water flowing northeast a few miles seaward from the reef, has been well known. Coral ecologist Walter Jaap reiterated that link in his comprehensive 1984 study of the reef: "The significance of the Florida Current cannot be overestimated when considering coral reef existence off southeast Florida."[19]

The sheer volume of the flow[20] makes the Florida Current impossible to ignore.

But as everyone who has visited the reef knows, there is another body of water, on the landward side of the reef, called Hawk Channel. Although it is relatively shallow and slow-moving compared to the magnificent Florida Current, Hawk Channel may be having a profound negative effect on the reefs. To be more precise, what's *in* Hawk Channel may be killing the reefs.

In 1987, Lapointe conducted a study with the soporific title "The effects of on-site sewage disposal systems on nutrient relations of groundwaters and nearshore surface waters of the Florida Keys."[21] In a perfect world, Lapointe's findings would have received more media attention than all the "SHIP HITS CORAL REEF!" stories that followed. But this isn't a perfect world, and Lapointe's study caused alarm—and controversy—only in scientific circles. What Lapointe found was that nutrients from human waste were entering the waters of Hawk Channel from 24,000 septic tanks, 5,000 illegal cesspits, and 200 water treatment facilities scattered throughout the Keys.[22] The introduction of seven hundred tons of nutrients annually[23] into the Keys' water fed algae, which, alleged Lapointe, smothered corals.

Critics immediately attacked Lapointe's conclusions. Some of the criticism was motivated by economic self-interest. Local developers (known as the Concrete Coalition) had been selling investors on the area's "gin-clear waters" for decades. Lapointe's findings contradicted these claims. "These waters are not pristine," he wrote, "but in fact stressed by nutrient loading."[24] Of course, the more successful developers were in attracting people to the Keys, the worse the sewage problem became. And they came in droves. The permanent population grew from 14,000 in 1940[25] to 50,000 in 1960[26] to 80,000 today.[27] That doesn't even include the 2.5 million people who vacation in the Keys each year.[28] While privately seething with anger, the economic development crowd attempted to laugh off Lapointe's charges by dismissing him as "Dr. Sewage."

But Lapointe's detractors extended beyond the Concrete Coalition. Respected scientists such as Alina Szmant, of the University of Miami's Rosenstiel School of Marine and Atmospheric Science, also questioned his conclusion. Szmant believed that Lapointe had demonstrated nutrient enrichment in nearshore waters, but she didn't see a direct link between that and problems on the reef. In her own studies, Szmant didn't detect elevated nutrient levels out on the reef tract.

"There is little justification to blame observed declines in coral cover on Florida coral reefs on anthropogenic nutrient enrichment," she wrote in a journal article, "a claim which is being made by environmental groups and the media."[29]

A researcher in the West Indies posted a question on an Internet discussion group devoted to coral health asking if anyone could supply figures at which nutrients are known to be harmful to corals. Szmant shot back her own posting about the dangers of looking for "simple answers" to complex questions.

"Please do not be mesmerized [by] the few investigators that are going around preaching for 'threshold' concepts," warned Szmant.

Lapointe has argued strenuously that such thresholds *do* exist.

When I ask Lapointe about these criticisms of his work, he immediately tears another bite out of his sandwich and grinds it furiously between his teeth, as if it were one of Alina Szmant's legs.

"Look," he says, after taking a gulp of sweet iced tea, "you simply can't hope to monitor nutrient levels as an indicator of coral health. You almost *have* to rely on bioindicators"—such as coral diseases or algae growth.

And even Szmant has some sympathy for that viewpoint. At a 1995 symposium on monitoring coral reef health, she noted that nutrients can be removed from the water column very quickly by photosynthesizers, or they can "hide" in sediments. She cited the problems of Kaneohe Bay, Hawaii, where algae had taken over corals "long before high water column nutrient concentrations were evident."[30]

But Szmant still wasn't convinced by Lapointe's thesis. What he called "bioindicators" of anthropogenic nutrient enrichment in the Keys, she saw as individual problems that could be caused by any number of conditions. Algae growth could be stimulated by nutrient upwellings from the Florida Current. Or it could be the continuing result of the *Diadema* die-off and overfishing. Or the coral could have died from something else first, perhaps bleaching, and *then* sprouted algae. Hard to say.

Lapointe, who thrives on controversy, plunged ahead.

"Sewage isn't the only threat to the reefs," he explains. "I started looking around and found what may be an even greater problem."

Lapointe backed up even further, focusing on the water on the *other* side of the Keys, an area that is known officially as Florida Bay, but referred to by locals simply as "the backcountry."

For every thousand sunburned tourists who make the drive down U.S. 1 along the backbone of the Keys, perhaps only one ever thinks about entering the backcountry. The only time most of them even notice that the area exists is at sunset, and then it is merely the setting for the colorful phenomenon that has been called the greatest show in the Keys. At all other times Florida Bay appears to be just a huge, flat expanse of shallow water dotted with countless mosquito-infested mangrove islands separating the Keys from the Everglades to the north. Boring.

Locals have always secretly treasured it, as much for the solitude and dreamlike atmosphere that pervades that sun-splashed realm of quiet waters as for the fisheries there. I've paddled a kayak around the backcountry many times, following lazy lemon sharks and manta rays as they skimmed over turtle-grass beds, and drifting within a few feet of herons preoccupied with finding a meal in the shallow waters. Physicists tell us that time and space do not really exist, and in the backcountry of the Keys you know, deep down in your gut, that they are right. The feeling that inhabits that place is like that of no other spot on Earth. The backcountry is the Keys' other, unknown, paradise.

At least it used to be.

If the Florida reef tract is in critical condition, then large sections of the backcountry may be terminally ill. The symptoms are numerous and diverse. Nearly seventy thousand acres of the seagrass meadows covering the bottom of the bay have died in recent years.[31] A massive *rhizoselenia* algae bloom of more than a hundred square miles has been dubbed the "Dead Zone" by fishermen who have had to abandon the area.[32] Tens of thousands of sponges, representing nearly every species found in the area, have died in the past few years, with mortality rates over 90 percent in many sites.[33]

To Lapointe, the manifold symptoms pointed to a single problem: nutrient loading.

"In the late 1980s I was told that there were no nutrients in Florida Bay," says Lapointe. "So I started sampling. Guess what I found? Alarmingly high levels of nutrients in Florida Bay."

And according to Lapointe, the nutrients wreaking havoc in Florida Bay flow out to Hawk Channel and onto the reefs, causing algae growth and coral diseases.

The obvious question is, How are the nutrients getting into Florida Bay in the first place? That's easy, says Lapointe. Nitrogen and phosphorus come from the massive sugarcane fields and vegetable farms south of Lake Okeechobee, more than a hundred miles to the north of the bay. Massive algae blooms on Okeechobee itself, and the proliferation of nutrient-loving cattails throughout the Everglades, seem to suggest that Lapointe is right. But other scientists studying the problem say that the decomposition of large amounts of sea grass could be the cause, not the symptom, of increased nutrients in the bay.

Ironically, argues Lapointe, the government's decision in 1991 to save the Everglades by restoring the historical flow of fresh water to that area (water that had been diverted fifty years ago to provide drinking water to the Miami area) may prove to be the greatest threat to the reef yet. The change, he says, would flood the already overloaded Bay with nutrient-rich waters. Predictably, others disagree. They maintain that the influx of fresh water is just what Florida Bay needs. According to this point of view, the seagrass die-offs, the great spreading blooms of algae, and the sponge mortality are caused by increased salinity, the result of diversion of fresh water.

It's hard for many to follow Lapointe's continual backstepping, blaming problems in the Florida reef tract on factors that are ever more remote. Yet reefs, by their coastal nature, are largely at the mercy of whatever material rivers carry to them from the land,

and there is no reason to believe that the Keys are any more immune to land-based pollutants and sediments than are the reefs in Indonesia and the Philippines. Even in Australia, some are arguing that the "lagoon," the body of water separating the mainland from the twelve-hundred-mile-long Great Barrier Reef, is experiencing the same problems that Lapointe describes. Peter Bell, a professor of chemical engineering at the University of Queensland and Lapointe's counterpart there, claims that parts of the lagoon have already turned eutrophic.[34] The increased flow of nutrients from agriculture and sewage is likely responsible for the death of some inner reefs, Bell argues, and if not checked, it could eventually threaten outer reefs as well.[35]

And that, according to Lapointe, is what's already happening here.

"It's clear that nutrients are affecting the reef from Long Key [in the north] all the way down to Key West," he says with typical certitude. "When you see half the coral die on a reef, you don't need a scientist to tell you what's going on."

Because he is so frequently quoted in the press, Lapointe has been derided as a publicity seeker. It's true that he doesn't avoid the media as some other scientists do, and that he uses the press as a pulpit from which to preach, a practice that makes many scientists squeamish. But Lapointe offers no apologies for any of this. Quite the contrary.

"My work *is* provocative," he admits. "If we're going to preserve and protect these reefs, we *have* to be."

Lapointe's is clearly a Gordian-knot solution, and it has all the advantages and disadvantages that come with that boldly narrow approach. He stays focused on the role of nutrient loading and doesn't muddy the waters (so to speak) with what he believes are minor issues. More trivial pursuits. There is only one trouble with this strategy: Lapointe had better be right, because in the tangled ball of problems he has so elegantly severed with a single stroke, there may lie other scientific knots, questions that could be vital to the reef's health.

In Australia, I ask Tony Larkum what he thinks about Lapointe and his theories. I half expect that the soft-spoken and ever-proper Larkum will arch an eyebrow and say something noncommittal but damning in a gentlemanly way, such as, "Ah, Lapointe. Interesting fellow."

Instead, Larkum looks at me unflinchingly and says gravely, "I'm afraid Brian may be more right than wrong."

❄

Billy Causey used to have the luxury of Lapointe's singleminded thinking. That was back in the old days, long before he became director of the Florida Keys National Marine Sanctuary (FKNMS). Billy, as he is known throughout the Keys, was a typical Conch (as natives and longtime residents are known), that is to say: as suspicious of big government as your average Montana Freeman, but too laid back to consider such lunatic ventures as building armed bunkers. Back then, Billy was just one of many tropical-fish collectors plying the waters of the middle Keys. When the government wanted to create a National Marine Sanctuary at Looe Key in 1981, Billy was one of the most vocal opponents of the plan. Mostly, he was afraid that a sanctuary would attract more people to the already heavily used reef. Everyone, including Billy, was stunned when the sanctuary was created and the government put him in charge of it. Whoever chose him must have been thinking of Lyndon Johnson's old vulgarism, that it's better to have your enemy on the inside of the tent pissing out, than on the outside pissing in. And the plan worked. Not only did his presence mollify many critics, but he became the very best kind of manager, able to enforce protective ordinances because they were *his* waters, too.

Federal officials were counting on the same thing happening when they appointed Billy head of the FKNMS. That may yet turn out to be the case. So far, however, the plan has backfired. Instead of the sanctuary gaining credibility because of him, in the eyes of many Conchs, Billy has become one of *them:* just another bureaucrat. That bothers him, but even after fifteen years on the "big government" payroll, there is enough of the Conch in Billy to enjoy the tussles.

"I was hung in effigy twice in one year," he says, with a note of perverse pride in his voice.

Six years after the creation of the FKNMS, Billy is pitching the concept to yet another group. This time it's to a small class taking a summer intensive college course on coral reef ecology at Pigeon Key, a tiny research station a third of the way down Seven Mile Bridge. I am a member of the class, the old man among college undergraduates.

Billy needs all the allies he can muster. Although the sanctuary was officially created years ago, the management plan still hasn't been approved, and a non-binding referendum on the proposed plan will be held in a few months. At the very least, Billy wants a majority of voters to approve the plan—he'd be happy with 50.1 percent—and he's here stumping for it.

In a land that only half-jokingly seceded from the Union a few years ago (to become the Conch Republic), the FKNMS is by definition a hard sell. There is a long history here of fiercely guarded independence. During the Civil War, a Key West resident is said to have told a Union officer stationed on the island, "At least we ain't got much to secede from 'cause we never really joined up."[36]

In many ways, not much has changed. Some adversaries of the sanctuary reject the "government intrusion" and liken it to "a system of total central command and control like the failed ones of Europe."[37] From Key Largo to Key West, "Say No to NOAA" bumper stickers and billboards are as common as coconuts.

Undaunted, Billy Causey presses his case.

"In the 1970s, average visibility was around a hundred feet out on the reef," he tells us. "Today it's thirty or forty feet—on a good day."

There is no question about that. I remember snorkeling in the Keys back in the 1970s. After the wind had laid down for two or three days, and the sediment had settled out, the water was so clear that it was like looking through air. Fish appeared to fly like birds. Your own body seemed released from the shackles of gravity, as you floated gently through liquid air. On every trip I've made back over the years, the water has grown increasingly more turbid. Now, when our Pigeon Key class dives on the reef, it's astonishing how much the visibility has diminished. If you remember what it was once like, you can actually feel claustrophobic diving on the reef today.

Billy explains that the management plan is based on a concept new to American marine sanctuaries: zoning. Of course, the idea is hardly new elsewhere. Australia's Great Barrier Reef Marine Park Authority has successfully used just such a system for two decades, allowing certain activities, such as commercial fishing, in some areas while banning it in others. And zoning is the key to the Apo reserve in the Philippines, not to mention many other areas around the world. The United States is coming to the concept of zoning in marine parks belatedly (not to mention kicking and screaming).

Mining and oil drilling within the 3,674-square-mile FKNMS were banned when the sanctuary was created in 1990, and a corridor around the reef was designated an "area to be avoided" by large vessels like the ones that ran aground in 1989. Aside from these changes, traditional uses will continue with little or no changes within most of the sanctuary. What is most striking about the plan, given all the controversy surrounding it, is how anemic it really is.

DeeVon Quirolo, project director and cofounder of the environmental group Reef Relief, charges that too many compromises were made in crafting a management plan she calls "a halfhearted job done poorly."[38]

She has a good point. Take fishing preserves as an example. When plans for the sanctuary were first being debated, one contingent wanted fishing banned in 25 persent of the FKNMS.[39] Jim Bohnsack, a research fisheries biologist with the National Marine Fisheries Service, and an adviser to the sanctuary, suggested a minimum of between 10 and 15 percent.[40] By March 1995, that "ecological preservation zone" had been whittled down to a mere two hundred square miles, or 5 percent of the sanctuary, spread out over three locations.[41]

Bohnsack, a realist, shrugged. "Five percent would be acceptable."[42]

When the final management plan was issued in 1996, two of the three preservation zones had been deleted. All that remained was a single no-fishing zone, covering eleven square miles, or one-third of one percent of the "sanctuary" total.[43]

At a presentation on the FKNMS given at the International Coral Reef Symposium in Panama, one fisheries biologist turned to a colleague and said with a sneer, "How the hell can you call something a *sanctuary* when you can take fish from virtually the whole thing?"

Regardless of how much the plan had been watered down, and despite Billy Causey's tireless stumping for the cause, in November 1996, 54.4 percent of voters rejected the proposition.[44] It's anyone's guess what will happen next.

❄

It is deep into a sweltering August night, and I'm unable to sleep. I hike up onto old U.S. 1, to where the original Seven Mile Bridge passes over Pigeon Key, spread out a towel smack in the center of the old road (which is closed to traffic), and lie down. We are two miles out from the mainland lights, and the stars sparkle above like phosphorescence in the water. Occasionally a shooting star streaks silently across the sky.

Of course, I am thinking about the reef—only five miles south across the dark waters of Hawk Channel. As I've done countless times during my travels and between interviews, I'm weighing the case for optimism versus pessimism, for this reef tract in particular (since it is the one I knew first and love the best), but also for reefs in general. Staring into the night sky, another braid suggests itself: I realize that whenever I consider reefs, feelings of hope and despair are intertwined. Which one, I wonder now, is more realistic? Even scientists disagree. Most are loath to use *either* word, for these are emotional terms, not scientific ones. But often, as you listen to scientists, you can hear the humanity breaking through the veneer of their scientific objectivity.

I recall the sober words of French scientist Bernard Salvat, as he addressed a workshop held the year before.

"To save the reefs for future generations will be very, very difficult," he told the scientists gathered in the Philippines. "Past and present situations do not allow us to be very optimistic. We have to be realists in order to act correctly and not to dream."[45]

He is right: coral reefs have never been so imperiled.

From global warming to sediment and nutrient loading, to overfishing and destructive fishing practices—there seem to be few areas of the tropical world immune to reef degradation.

But I also recall the words of Gregor Hodgson at Hong Kong University. "Everyone is tired of gloom and doom," he wrote in a gentle rebuke to a note of my own, confessing pessimism. "Coral reefs are much tougher than most scientists believe," he went on. "While degradation is rampant, total destruction such as occurred in Jakarta Bay is rare. And the discussion level about coral reefs has never been higher in governments around the world."

He is right: coral reefs have never commanded so much attention.

Standing on the margin of the Great Barrier Reef in late 1996, President Bill Clinton extolled Australia's commitment to reef science and management. "Today, with your knowledge and leadership," he told a cheering crowd, "we are seeing to it that the world's reefs make it into the next century safe and secure."[46]

The popular press covers reef issues with greater nuance and consistency than in the past. Also in 1996, *Time* magazine ran a cover story about "Treasures of the Seas" highlighting coral reefs.[47]

More than fifty organizations around the world joined together to promote 1997 as "The International Year of the Reef," with a slew of public educational programs, workshops for park managers, scientists and recreational divers to collaborate, and "Reef Check," a worldwide effort to assess conditions on reefs. Scientists know more then ever before about the complex workings of this marvelous ecosystem and its relationship to mangroves and seagrass beds.

But is it enough? Knowledge is a poor substitute for wisdom, and surely what we lack most isn't information (which, at any rate, will always be incomplete). What we need is the wisdom to cast our lot with the rest of the natural world, to understand our place within this enchanted braid of life and, as Rachel Carson put it, to master ourselves. We have given ourselves the grand name *Homo sapiens*, "wise man." But that's mere bravado. Until we understand who and what and where we are, we will remain a distant second best: *Homo versutus*, "clever man."

Hope or despair—which one is more appropriate?

Staring into the cavernous sky, I listen for answers in the night. But of course there are no answers, just the sound of the waves lapping on the rocks below.

❄

The next day the sun is on the sea, and the gloomy thoughts of the previous night dissolve in the dazzling light, their remains swept away on a warm Caribbean breeze. Hope has replaced despair, for today my parents drive down the Keys, bringing my wife, Mary, and toddler son, Liam, for a short visit. When they finally arrive, I realize that I hadn't allowed myself to feel the depth of my loneliness until I had my family around me again. I take a break from the classroom to show them around the tiny island. We tour the lab where fragments of *Acropora palmata* grow in large, bubbling tanks. We walk out onto the pier, on the Florida Bay side of the key, and look into the water where I have spent several hours snorkeling, learning to identify corals, sea grasses, algae, and fishes. I get down on my hands and knees, holding my young son around the waist so he can gaze into the water and see Sam, the sleek barracuda who lives under the dock like a troll in a children's story.

"Do you see him?" I ask.

Liam points down and cries, "Fish!"

My parents and Mary wait in the shade, while Liam and I work our way around the island.

"Keep his hat on!" Mary yells.

It suddenly seems vital to show him everything, somehow to communicate to him how important this place and its inhabitants are. But of course it's futile. He knows so few words. And besides, my knowledge is woefully inadequate to the task. But he seems to be enjoying himself as he rides on my hip, and so I point to the creatures caught in tide pools and to plants and animals a bit farther out, reciting the names of the ones I know as if chanting a liturgy.

Chitons. *Truncatella scalaris.* Grapsoid crabs. Sipunculid worms. Palythoa. Sea stars. *Thalassia.*

Finally we arrive on the dark, weathered rocks that mark the southern tip of Pigeon Key. My arms are laced around Liam's back, and his tiny hands grip my shoulders as the tide hisses into crevices below. I point to the horizon, where a lighthouse marks Sombrero Reef.

"Out there is the reef."

He looks, blinking silently beneath the brim of his hat into the fierce sunlight.

Embracing my son, I recall how two decades ago I used to stare out beyond the reef and imagine that the next landmass was Africa. Pointing east of the lighthouse, I tell Liam, "And that way is Africa."

He looks over at me for a second, and smiles the way he often does when I've said something silly. Then he gazes back out to sea. He turns solemnly silent, as only a two-year-old can. It's almost as if he knows that it's not just Africa lying out beyond the vast glimmering waters, but his future.

Notes

1 "Who Has Known the Ocean?"

1. Carson 1937:322.
2. Couper 1983:7.
3. MacLeish 1989:16.
4. Reaka-Kudla 1997:86.
5. Abercrombie et al. 1990:406.
6. Broad 1997:40.
7. Waller 1996:32.
8. Waller 1996:33.
9. Thurman 1987:8.
10. Bult 1996:1066.
11. Reaka-Kudla 1996:88.
12. Birkeland 1997:303.
13. *Science* 2 February 1996:597.
14. Ian Whittington, author interview, 6 November 1996.
15. Reaka-Kudla 1997:87.
16. Reaka-Kudla 1997:87.
17. Reaka-Kudla 1997:93.
18. Jackson 1995:6.
19. Reaka-Kudla 1997:88.
20. Gardiner 1931:1.
21. Wallace, Alfred R. 1869:226.
22. Darwin 1988:397.
23. Darwin 1988:403.
24. Gardiner 1931:1.
25. Dubinsky 1990:345.
26. Hay 1997.
27. Jameson et al. 1995:3.

2 Animal, Mineral, Vegetable

1. Yonge 1931:3.
2. Veron 1986:46.

3. Veron 1986:45.
4. Birkeland 1997:98.
5. Putman and Wratten 1984:316–17.
6. Fagerstrom 1987:22.
7. Sorokin 1993:1.
8. Heywood 1995:381.
9. Sorokin 1993:5.
10. Heywood 1995:381.
11. Maragos et al. 1996:97.
12. Dana 1843:130.
13. Veron 1995:96.
14. McManus 1988:189.
15. Eakin et al., in press.
16. Birkeland 1996:5.
17. NOAA 1995:abstract.
18. Wilkinson 1994:9.
19. McManus 1988:189.
20. Maragos et al. 1996:87.
21. Peter Sale, author interview, 25 June 1996.
22. Birkeland 1996:5.
23. Fairbanks, in press.
24. Peter Sale, E-mail to the author, 13 June 1996.

3 Darwin in Paradise

1. Francis Darwin 1877:70.
2. Browne 1995:13.
3. Browne 1995:81.
4. Browne 1995:81.
5. Browne 1995:317.
6. Darwin 1988:10.
7. Darwin 1988:218.
8. Darwin 1988:218.
9. Browne 1995:186.
10. Browne 1995:362.
11. Darwin 1988:218.
12. Darwin 1988:348.
13. Darwin 1988:351.
14. Browne 1995:318.
15. Darwin 1988:393.
16. Darwin 1984:6.
17. Darwin 1988:399.

18. Darwin 1988:398–99.
19. Darwin 1984:41.
20. Veron 1986:5.
21. Francis Darwin 1887:362.
22. Wells, vol. 3, 1988:308.
23. Kaplan 1982:99.
24. Ladd et al. 1953:2277.
25. Ladd et al. 1953:2277.
26. Wells 1988:216.
27. Ladd and Schlanger 1960:866–67.
28. Ladd et al. 1953:2259.
29. Ladd and Schlanger 1960:863.
30. Ladd et al. 1953:2264.
31. Ladd et al. 1953:2264.
32. Ladd et al. 1953:2265.
33. Ladd et al. 1953:2265.
34. Ladd et al. 1953:2267.
35. Ladd et al. 1953:2258.
36. Ladd et al. 1953:2266.
37. Ladd and Schlanger 1960:899.
38. Ladd et al. 1953:2276.
39. All of the information on the Mike explosion comes from Rhodes, 1995:482–512.
40. Shepley and Clair 1954:151.

4 The Rise of Corals

1. Veron 1986:624.
2. John Wells 1963:950.
3. Veron 1995:111.
4. Couper 1983:34–35.
5. Dubinsky 1990:5.
6. Dubinsky 1990:5.
7. Sorokin 1995:8.
8. Wilson 1992:30.
9. Dubinsky 1990:5.
10. Sorokin 1995:8.
11. Birkeland 1997:15.
12. Veron 1986:624.
13. Birkeland 1997:31.
14. Waller 1996:187.
15. Fagerstrom 1987:280.

16. Veron 1995:242.
17. Veron 1986:126.
18. Jackson 1995:6.
19. Jackson 1995:6.
20. Waller 1996:88.
21. Sandra Romano, E-mail to the author, 14 February 1997.
22. For more on these adaptations, see Veron, 1986:126, and Veron, 1995:242–43.
23. Frith and Frith 1992:12.
24. Peter Sale, E-mail to the author, 13 June 1996.
25. Garraty and Gay 1972:72.
26. Peter Sale, author interview, 25 June 1996.
27. Guilcher 1988:50.
28. Veron 1995:123.
29. Veron 1995:122.
30. Birkeland 1997:377.
31. Birkeland 1997:39.
32. Birkeland 1997:49.
33. Fagerstrom 1987:71.
34. Bird 1993:129.
35. Birkeland 1997:39.

5 The Heart of Lightness

1. Veron 1995:158–159.
2. Wallace 1869:185.
3. Wallace 1869:185–218.
4. Whitten 1987:59.
5. Quammen 1996:57.
6. Quoted in Whitten 1987:217.
7. Odum and Odum 1955:291–320.
8. Kaplan 1982:103.
9. Odum 1975:37.
10. Birkeland 1997:54.
11. Bold and Wynne 1985:583.
12. Whitten 1987:221.
13. Sorokin 1995:15.
14. Carson 1955:227.
15. Waller 1996:180.
16. Waller 1996:116.
17. Guilcher 1988:29.
18. Hopley 1989:4.

6 The Outer Strands

1. Eugene Odum 1975:39.
2. Jackson 1995:7.
3. Phillips and Meñez 1988:5.
4. Phillips and McRoy 1980:93.
5. Chiappone et al. 1996:55.
6. Phillips and Meñez 1988:5.
7. Phillips and Meñez 1988:9.
8. Cited by Simon Wilkinson, *Quarterly News* 4, no. 3 (Winter 1996), Mangrove Action Project, 9.
9. Darwin 1984:432.
10. Tomlinson 1986:7.
11. Heywood 1995:390.
12. Lurie 1988:228.
13. Carson 1955:207.
14. Waller 1996:83.
15. "On the Status of Mangrove Forests Worldwide," *Quarterly News* 2, no. 2 (Summer 1996), Mangrove Action Project, 5.
16. Gato 1991:34.
17. Tomlinson 1986:7.
18. Tomlinson 1986:109.
19. Wilkinson 1994:51.
20. Ogden and Gladfelter 1983:7.

7 A Song of Love and Death

1. Dubinsky 1990:133.
2. Vaughan and Wells 1943:41–42.
3. Duerden 1905:94.
4. Thynne, cited in Dubinsky 1990:133.
5. Szmant 1996.
6. Jamie Oliver, author interview, 10 November 1996.
7. Stimson 1978:182.
8. Earle 1995:69.
9. Szmant-Froelich et al. 1980.
10. Dubinsky 1990:134.
11. Jamie Oliver, author interview, 10 March 1997.
12. Dubinsky 1990:166.
13. Harrison et al. 1984:1188.
14. Harrison et al. 1984:1188.
15. Dubinsky 1990:134.

16. Birkeland 1997:177.
17. Willis 1997.
18. Dubinsky 1990:184.
19. Birkeland 1997:190.
20. Birkeland 1997:190.
21. Grigg 1981:16.
22. *Encyclopaedia Britannica*, vol. 17, 1990:581.
23. Grigg 1981:17.
24. Edinger and Risk 1995:200.
25. Sammarco, in press.
26. Lang and Chornesky 1990:212.
27. Sorokin 1995:283.
28. Lang and Chornesky 1990:212.
29. Lang and Chornesky 1990:218.
30. Sammarco et al. 1983:173–78.
31. Lang and Chornesky 1990:218.

8 Fish Stories

1. Vaccari and Jackson 1995:70.
2. Bellwood 1996:11.
3. Landini and Sorbini 1996:105.
4. Birkeland 1996:303.
5. Sale 1991:5.
6. Sale 1991:61.
7. Sale 1991:31.
8. Sale 1991:49.
9. Sale 1991:76.
10. *New York Times*, "Man Who Swallows Fish Dies as Joke Goes Awry," 1 April 1997.
11. McCosker 1977:400–401.
12. Sale 1991:479.
13. See Kaplan 1982:246–48; Sale 1991:590; Roessler 1977:98.
14. Waller 1996:214.
15. Birkeland 1996:260.
16. Thresher 1984:370.
17. Thresher 1984:389.
18. Dubinsky 1990:354.
19. Thresher 1984:37.
20. Thresher 1984:343.
21. Sale 1991:332.
22. Peter Sale, E-mail to the author, 16 September 1996.

23. Warner et al. 1975:636.
24. Thresher 1984:216.
25. Thresher 1984:391.
26. Thresher 1984:391.
27. Warner et al. 1975:636.
28. Warner and Swearer 1991:199–204.
29. Warner and Swearer 1991:200.
30. Warner and Swearer 1991:199.
31. Peter Sale, E-mail to the author, 16 September 1996.
32. Thresher 1984:77.
33. Sale 1991:184–85.
34. Sale 1991:184–85.
35. Sale 1991:183.
36. Sale 1991:186.
37. Sale 1991:264–65.
38. Sale 1991:48.
39. Sale 1991:97.
40. Sale 1991:97.
41. Sorokin 1995:232.
42. Sorokin 1995:234.
43. Wilkinson 1994:11.
44. Vine 1974:131.
45. Vine 1974:132.
46. Lassuy 1980:311.
47. Sale 1991:81

9 Neither Brethren nor Underlings

1. Darwin 1988:396.
2. Colin Limpus, fax to the author, 6 January 1997, and 9 January 1997.
3. Nancy FitzSimmons, E-mail to the author, 20 December 1996.
4. Lutz and Musick 1997:126.
5. Lutz and Musick 1997:56.
6. Lutz and Musick 1997:38.
7. Jackson, in press.
8. Lutz and Musick 1997:58.
9. Lutz and Musick 1997:221.
10. Steve Grenard, E-mail to the author, 23 April 1997.
11. Lutz and Musick 1997:142.
12. Lutz and Musick 1997:preface.
13. Lutz and Musick 1997:279.

14. Lutz and Musick 1997:347.
15. Jeanette Wyneken, E-mail to the author, 5 April 1997.
16. Nancy FitzSimmons, E-mail to the author, 20 December 1996.
17. Lutz and Musick 1997:59.
18. Lutz and Musick 1997:109.
19. Lutz and Musick 1997:65.
20. Judy Simmons, E-mail to the author, 31 March 1997.
21. Sale 1991:222.
22. Lutz and Musick 1997:117.
23. Lutz and Musick 1997:129.
24. Morreale et al. 1996:319.
25. Lutz and Musick 1997:200–201.
26. Humann 1992:108.
27. Lutz and Musick 1997:201.
28. Jackson, in press.
29. Jackson, in press.
30. Lutz and Musick 1997:212.
31. Lutz and Musick 1997:38.
32. Lutz and Musick 1997:58.
33. Lutz and Musick 1997:57.
34. Lutz and Musick 1997:57.
35. Lutz and Musick 1997:58.
36. Beston 1928:25.
37. National Research Counsil 1990:182.

10 The Jakarta Scenario

1. Umbgrove 1939:10.
2. Ginsburg 1994:308.
3. Robin Harger, author interview, 25 June 1996.
4. Ginsburg 1994:308.
5. Terry Done, author interview, 11 November 1996.
6. Dalton 1995:191.
7. S. M. Evans et al. 1995:109.
8. World Bank 1994:xiv.
9. Laber 1997:44.
10. Laber 1997:40.
11. World Bank 1994:xi.
12. Hungspreugs 1988:181.
13. World Bank 1994:70.
14. Dalton 1995:191.
15. World Bank 1994:xx.

16. Information provided author by Evan Edinger from samples taken 30 July–13 August 1996.
17. Cliff Davidson, E-mail to the author, 5 December 1996.
18. World Bank 1994:81.
19. Evans et al. 1995:109–14.
20. World Bank 1994:91.
21. World Bank 1994:70.
22. World Bank 1994:71.
23. World Bank 1994:69.
24. Cesar 1996:59.
25. Hungspreugs 1988:178, and Robin Harger, author interview, 25 June 1996.
26. Cesar 1996:59.
27. Hungspreugs 1988:181.
28. UNESCO 1986:123.
29. UNESCO 1996:22.
30. Umbgrove 1947:736.
31. Ongkosongo 1986:134.
32. Ongkosongo 1986:134.
33. Evan Edinger, author interview, 21 November 1996.
34. Wilkinson 1994:44.
35. Djoekardi 1995:2.
36. Mike Risk, author interview, 22 November 1996.
37. Birkeland 1997:1.
38. Wilkinson 1994:10.
39. Wilkinson 1994:11.
40. Voice of America broadcast, 21 April 1997.
41. McManus 1997.

11 "Either We Go Deep or We Starve"

1. Willoughby et al. 1996:vi–13.
2. Evan Edinger, author interview, 20 November 1996.
3. Willem Moka, author interview, 16 November 1996.
4. McAllister and Ansula 1993:43.
5. Erdmann 1995:4.
6. Wells 1988, vol. 2:122.
7. Moll 1983:27.
8. McAllister and Ansula 1993:43.
9. Erdmann, in press.
10. Willoughby et al. 1996:vi–12.

11. Rodney Salm, E-mail to the author, 6 November 1996.
12. Wells 1988.
13. ReefBase 1996.
14. Wells 1988, vol. 2:365.
15. Rodney Salm, E-mail to the author, 6 November 1996.
16. McAllister and Ansula 1993:43.
17. McManus 1988:191.
18. Mike Risk, author interview, 22 November 1996.
19. Rodney Salm, E-mail to the author, 6 November 1996.
20. Umbgrove 1947:735.
21. ReefBase 1996.
22. Soekarno 1989:26.
23. Sergio Cotta, author interview, 18 November 1996.
24. Evan Edinger, author interview, 20 November 1996.
25. Erdmann 1995:5.
26. Erdmann 1995:5.
27. Marie-Trees Meereboer, author interview, 16 November 1996.
28. *Life Reef Fish Information Bulletin*, March 1996:3.
29. *South China Morning Post*, 28 November 1996.
30. *South China Morning Post*, 28 November 1996.
31. Erdmann and Pet-Soede 1996:2.
32. Johannes and Riepen 1995:14.
33. Johannes and Riepen 1995:28.
34. Cesar, 1996:3.
35. Calum Roberts, E-mail to the author, 29 November 1996.
36. Johannes and Riepen 1995:10.
37. Erdmann and Pet-Soede 1996:4.
38. Johannes and Riepen 1995:13.
39. Calum Roberts, E-mail to the author, 29 November 1996.
40. Johannes and Riepen 1995:18.
41. Johannes and Riepen 1995:18.
42. Michael Aw, E-mail posting, 3 December 1996.
43. Steve Oakley, E-mail posting, 9 December 1996.
44. Lutz and Musick 1997:17.

12 The Apo Scenario

1. McManus 1996:250.
2. McManus, in press.
3. Salvat 1995:12.
4. Chris Crossland, in press.
5. Alcala 1988:196.

6. Alcala 1988:196.
7. Savina and White 1986:111.
8. Alcala 1988:196.
9. Roberts and Polunin 1993:365.
10. Russ and Alcala 1994:11.
11. Vogt, in press.
12. Russ and Alcala 1996:5.
13. Russ and Alcala 1996:6–7.
14. Douglas Fenner, E-mail to the author, 22 November 1996.
15. Hinrichsen 1996:22.
16. Vogt, in press.
17. Sotto, Filipina, E-mail to the author, 6 February 1997.
18. Russ 1996:8.
19. Russ 1996:9.

13 Return to Oceanus

1. Thurman 1987:179.
2. Thurman 1987:178.
3. Thurman 1987:39.
4. Dario Sandrini, E-mail to the author, 10 November 1996.
5. Evan Edinger, author interview, 20 November 1996.
6. Wilkinson 1994:18.
7. Eric Mueller, author interview, 11 August 1996.
8. Ginsburg 1994:340.
9. Ginsburg 1994:340.
10. Birkeland 1997:371.
11. Johannes 1971:9.
12. Johannes 1971:9.
13. Ginsburg 1994:340.
14. Birkeland 1997:371.
15. Wells 1988:165.
16. Ginsburg 1994:341.
17. Ginsburg 1994:341.
18. Ginsburg 1994:343.
19. F. R. Fosberg, quoted in Johannes 1971:186.
20. Johannes 1971:186.
21. Johannes 1971:186.
22. Ginsburg 1994:343.
23. Birkeland 1997:376.
24. Wilson 1992:197.
25. Wilson 1992:198.

26. Kricher 1989:283.
27. Kricher 1989:166.
28. Wong and Ventocilla 1995:142.
29. Kricher 1989:329.
30. Steve Hendrix, author interview, 3 November 1997.
31. Wong and Ventocilla 1995:23.
32. Kricher 1989:67.
33. Dubinsky 1990:401.
34. Wong and Ventocilla 1995:29.
35. Kricher 1989:71.
36. Birkeland 1997:6.
37. Wong and Ventocilla 1995:63.
38. Wilson 1992:274.
39. Wilson 1992:275.
40. Dubinsky 1990:443–44.
41. Ginsburg 1994:vi.
42. Susan Wells, E-mail to the author, 20 May 1997.
43. Birkeland 1997:6.
44. Birkeland 1997:394.
45. Hodgson and Dixon 1988:8.
46. Wilson 1992:243.
47. Hodgson and Dixon 1988:1.
48. Hodgson and Dixon 1988:30.
49. Birkeland 1997:395.
50. Hodgson and Dixon 1988:37.
51. Hodgson and Dixon 1988:41.
52. Hodgson and Dixon 1988:63.
53. Hodgson and Dixon 1988:68.
54. Gregor Hodgson, E-mail to the author, 25 November 1996.
55. Gregor Hodgson, E-mail to the author, 25 November 1996.
56. Thomas Goreau, in press.

14 Disasters, Catastrophes, and Tragedies

1. Wilson 1992:19.
2. Wilson 1992:22.
3. Connell 1978:1303.
4. Erich Mueller, author interview, 14 August 1996.
5. Terres 1991:734.
6. Lessios et al. 1984:335.
7. Jackson, in press.
8. Jackson, in press.

9. Jackson, in press.
10. Kaplan 1982:19.
11. Humann 1992:286.
12. Hendrickson 1978:72.
13. Lessios et al. 1984:336.
14. Birkeland 1997:127.
15. Lessios 1988:372.
16. Lessios 1988:386.
17. Lessios 1988:373.
18. Lessios 1988:374.
19. Samarrai 1995:16.
20. Hughes 1994:1549.
21. Hughes 1994:1549–50.
22. Jackson, in press.
23. Jackson, in press.
24. Hughes 1994:1548.
25. Jackson, in press.
26. Jackson 1995:7.
27. Jackson, in press.
28. Jackson, in press.
29. Jackson, in press.
30. Roberts 1994:266.
31. NOAA 1994:6.
32. Couper 1983:44.
33. Fairbanks, in press.
34. NOAA 1994:2.
35. NOAA 1994:2.
36. NOAA 1994:2.
37. Glynn 1984:140.
38. Glynn 1991:176.
39. NOAA 1994:2.
40. NOAA 1994:3.
41. Glynn 1983:149.
42. Glynn 1983:149.
43. Meehan and Ostrander 1997:104.
44. *CRC Reef Research News* 1996:1.
45. Bunkley-Williams and Williams 1990:48–49.
46. Glynn 1991:177.
47. Glynn 1984:138.
48. *CRC Reef Research News* 1996:1.
49. Glynn 1984:142.
50. Glynn 1991:175.

51. Brown and Ogden 1993:70.
52. Glynn 1983:150.
53. Glynn 1983:150.
54. Glynn 1983:151.
55. Glynn 1991:177.
56. Glynn 1983:152.
57. Birkeland 1997:366.
58. Glynn 1984:143.
59. Bunkley-Williams and Williams 1990:52.
60. Goreau and Hayes 1994:177–78.
61. Bunkley-Williams and Williams 1990:48.
62. Bunkley-Williams and Williams 1990:52.
63. U.S. Congress, Senate, 1988.
64. U.S. Congress, Senate, 1988:10.
65. U.S. Congress, Senate, 1991:12.
66. Carnegie Mellon Internet paper, page 1.
67. Bunkley-Williams and Williams 1990:52.
68. U.S. Congress, Senate, 1991:16.
69. U.S. Congress, Senate, 1991:34.
70. Harper and Fullerton 1993:4.
71. Goreau and Hayes 1994:179.
72. Hayes and Strong 1996:2.
73. Brown, in press.
74. NOAA press release, "Coral Reef Bleaching Found in Belize for the First Time," 10 November 1995.
75. Goreau, in press.
76. Roger Uwate, E-mail posting, 23 September 1996.
77. Bruce Carlson, E-mail posting, 25 September 1996.
78. NOAA, Diagnostic Advisory, 10 October 1997.
79. McKibben 1989:172.

15 Unweaving the Outer Braids

1. Dalton 1995:181.
2. Barraclough and Finger-Stich 1996:iii–7.
3. Barraclough and Finger-Stich 1996:iii–7.
4. Evan Edinger, author interview, 20 November 1996.
5. Barraclough and Finger-Stich 1996:ii–2.
6. Barraclough and Finger-Stich 1996:ii–2.
7. Barraclough and Finger-Stich 1996:ii–3.
8. Chua 1992:101.
9. Barraclough and Finger-Stich 1996:i–2.

10. Jean-Luc de Kok, author interview, 14 November 1996.
11. Barraclough and Finger-Stich 1996:i–2.
12. Anthony Viner, author interview, 14 November 1996.
13. Wilkinson et al. 1994:216.
14. Wilkinson et al. 1994:216.
15. Wilkinson et al. 1994:235.
16. Wilkinson 1994:52.
17. Wilkinson et al. 1994:204.
18. Wilkinson 1994:52.
19. Alfredo Quarto, E-mail to the author, 3 October 1996.
20. John McManus, author interview, 25 November 1996.
21. Barraclough and Finger-Stich 1996:iv–9.
22. Barraclough and Finger-Stich 1996:iv–7.
23. Wilkinson 1994:107.
24. Ducker et al. 1977.
25. Hutomo, author interview, 22 November 1996.
26. Wilkinson et al. 1994:260.
27. Wilkinson 1994:107.
28. Tony Larkum, author interview, 2 November 1996.
29. Wilkinson et al. 1994:259.
30. Cooperative Research Center 1995:1.
31. Tony Larkum, author interview, 27 June 1996.
32. Wilkinson 1994:112.
33. Larkum and West 1990:66.
34. Larkum and West 1990:66.
35. Larkum and West 1990:61.
36. Larkum, E-mail to the author, 27 June 1997.
37. Wilkinson 1994:107.

16 Once More to the Keys

1. Derr 1989:310.
2. Pamphlet, Key West's Shipwreck Historeum, 1994.
3. Artman 1969:4.
4. Windhorn and Langley 1973:29.
5. Derr 1989:310.
6. Jaap 1984:4.
7. Gittings and Brights 1988:37.
8. *The Economist*, 9 January 1988:26.
9. Gittings 1991:2–5.
10. United Press International, 1 March 1990.
11. Gittings 1991(b):1–10.

12. U.S. Congress, House, 1990:1.
13. U.S. Congress, House, 1990:4.
14. U.S. Congress, *Statutes at Large*, 1991:104 Stat. 3089.
15. U.S. Congress, House, 1990:36.
16. U.S. Congress, House, 1990:29.
17. Erich Mueller, author interview, 9 August 1996.
18. Porter and Meier 1992:637.
19. Jaap 1984:1.
20. Jaap 1984:12.
21. Lapointe and O'Connell 1988.
22. Lapointe and Clark 1990:12.
23. *Reef Line*, Spring 1996:2.
24. Lapointe and O'Connell 1988:9.
25. Chiaponne et al., vol. 3, 1996:15.
26. Chiaponne et al., vol. 3, 1996:17.
27. Chiaponne et al., vol. 3, 1996:19.
28. *Palm Beach Post*, 24 February 1997.
29. Szmant and Forrester 1996:39.
30. Crosby et al. 1996:58.
31. Chiappone et al., vol. 2, 1996:34.
32. *Palm Beach Post*, 1 June 1993.
33. Chiappone et al., vol. 9, 1996:24.
34. Bell 1992.
35. Bell 1992:553.
36. Windhorn and Langley 1973:10.
37. *Palm Beach Post*, 24 February 1997.
38. *Palm Beach Post*, 24 February 1997.
39. DeeVon Quirolo, author interview, 28 June 1996.
40. Jim Bohnsack, author interview, 26 June 1996.
41. NOAA 1995:10.
42. Jim Bohnsack, author interview, 26 June 1996.
43. John Ogden, E-mail to the author, 14 January 1997.
44. *Dive Training*, January 1997:15.
45. Salvat 1995:13.
46. Remarks by U.S. President Bill Clinton, Port Douglas, Australia, 22 November 1996.
47. *Time*, Australian edition, 28 October 1996.

Bibliography

Abercrombie, Michael, et al. *The New Penguin Dictionary of Biology*. London: Penguin Books, 1990.

Alcala, Angel. "Effects of Marine Reserves on Coral Fish Abundance and Yields of Philippine Coral Reefs." *Ambio* 17, no. 3 (1988), 194–99.

Artman, L. P. *Key West History*. Self-published, 1969.

Barraclough, Solon, and Andrea Finger-Stich. "Some Ecological and Social Implications of Commercial Shrimp Farming in Asia." Geneva and London: United Nations Research Institute for Social Development and the World Wildlife Fund, 1996.

Bell, Peter. "Eutrophication and Coral Reefs—Some Examples in the Great Barrier Reef Lagoon." *Water Resources* 25 (1992), 553–68.

Bellwood, D. "The Eocene fishes of Monte Bolca: the earliest coral reef fish assemblage." *Coral Reefs* 15 (1996), 11–19.

Beston, Henry. *The Outermost House*. New York, Toronto: Rinehart & Co., 1949.

Bird, Eric. *Submerging Coasts*. New York: John Wiley & Sons, 1993.

Birkeland, Charles, ed. *Life and Death of Coral Reefs*. New York: Chapman & Hall, 1997.

Bold, Harold, and Michael Wynne. *Introduction to the Algae*. Englewood Cliffs, N.J.: Prentice Hall, 1985.

Broad, Robin, with John Cavanagh. *Plundering Paradise*. Berkeley: University of California Press, 1993.

Broad, William. *The Universe Below*. New York: Simon and Schuster, 1997.

Brown, Barbara. "Coral Bleaching: Causes and Consequence." In *The Proceedings of the Eighth International Coral Reef Symposium*. In press.

Brown, Barbara, and John Ogden. "Coral Bleaching." *Scientific American*, January 1993, 64–70.

Brown, Janet. *Charles Darwin: Voyaging*. Princeton, N.J.: Princeton University Press, 1995.

Bult, C., et al. "Complete genome sequence of the methanogenic archaoen, *Methanococcus jannaschii.*" *Science*, 23 August 1996, 1066.

Bunkley-Williams, Lucy, and Ernest Williams Jr. "Global Assaults on Coral Reefs." *Natural History*, April 1990, 47–54.

Carson, Rachel. "Undersea." *The Atlantic Monthly*, September 1937. 322.

———. *The Sea Around Us.* Oxford: Oxford University Press, 1950.

———. *The Edge of the Sea.* New York: Houghton Mifflin, 1955.

Cesar, Herman. "Economic Analysis of Indonesian Coral Reefs." Washington, D.C.: The World Bank, December 1996.

Chiappone, Mark, et al. "Oceanography and Shallow-water Processes of the Florida Keys and Florida Bay." In *Site Characterization for the Florida Keys National Marine Sanctuary and Environs, vol. 2.* Zenda, Wisc.: The Preserver, 1996.

Chua, Thia-Eng. "Coastal Aquaculture Development and the Environment." *Marine Pollution Bulletin* 25 (1992), 1–4, 98–103.

Connell, Joseph. "Diversity in Tropical Rain Forests and Coral Reefs." *Science*, 24 March 1978, 1302–9.

Cooperative Research Center. "Coral Bleaching." *CRC Reef Research News*, December 1996, 1–2.

———. "Vast Seagrass Meadows Found in the Reef Depths." *CRC Reef Research News*, October 1995, 1–2.

Couper, Alastair. *The Times Atlas of the Oceans.* New York: Van Nostrand Reinhold, 1983.

Crosby, M. P., et al., eds. *A Coral Reef Symposium on Practical, Reliable, Low Cost Monitoring Methods for Assessing the Biota and Habitat Conditions of Coral Reefs*, 26–27 January 1995. Silver Springs, Md.: Office of Ocean and Coastal Resource Management, National Oceanic and Atmospheric Administration, 1996.

Crossland, Chris. "Making Reef Science Relevant: A Great Barrier Reef Case." In *Proceedings of the Eighth International Coral Reef Symposium.* In press.

Dalton, Bill. *The Indonesia Handbook.* Chico, Calif.: Moon Publications, 1995.

Dana, James. "On the Temperature limiting the Distribution of Corals." *American Journal of Science* 45 (1843), 130–31.

Darwin, Charles. *The Structure and Distribution of Coral Reefs.* 1842. Reprint, Tucson: University of Arizona Press, 1984.

———. *The Voyage of the Beagle.* 1839. Reprint, New York: New American Library, 1988.

Darwin, Francis, ed. *The Life and Letters of Charles Darwin.* London: John Murray, 1887.

Derr, Mark. *Some Kind of Paradise.* New York: William Morrow, 1989.

Djoekardi, Arie. "Urban Land Use Planning Policy in Indonesia." Seoul, South Korea: Workshop on Policy Measures for Changing Consumption Patterns, August 1995.

Dubinsky, Zvy, ed. *Coral Reefs: Ecosystems of the World,* vol. 25. Amsterdam: Elsevier, 1990.

Ducker, S. C., et al. "Biology of Australian seagrasses: the genus *Amphibolis* C. Agardh (Cymodoceaceae)." *Australian Journal of Botany* 25 (1977), 67–95.

Duerden, J. "Recent results on the morphology and development of coral polyps." *Smithsonian Miscellaneous Collection* 47 (1905), 93–111.

Eakin, Mark, et al. "Coral reef status around the world: where are we and where do we go from here?" In *Proceedings of the Eighth International Coral Reef Symposium.* In press.

Earle, Sylvia. *Sea Change: A Message of the Oceans.* New York: Fawcett Columbine, 1996.

Edinger, Evan, and Michael Risk. "Preferential survivorship of brooding corals in a regional extinction." *Paleobiology* 21 (1995), 200–219.

Ekachai, Sanitsuda. *Behind the Smile: Voices of Thailand.* Bangkok: 1994.

Erdmann, Mark. "An ABC Guide to Coral Reef Fisheries in Southwest Sulawesi, Indonesia." *Naga,* April 1995, 4–6.

———. "Destructive Fishing Practices in the Pulau Seribu Archipelago." *UNESCO Reports in Marine Science.* In press.

Erdmann, Mark, and Lida Pet-Soede. "How Fresh Is Too Fresh? The Live Reef Food Fish Trade in Eastern Indonesia." *Naga,* January 1996.

Evans, S. M., et al. "Domestic Waste and TBT Pollution in Coastal Areas of Ambon Island." *Marine Pollution Bulletin* 30, no. 2 (1995), 109–15.

Fagerstrom, J. A. *The Evolution of Reef Communities.* New York: John Wiley & Sons, 1987.

Fairbanks, Richard. "Extracting the Climate Record from Coral Reefs." In *Proceedings of the Eighth International Coral Reef Symposium.* In press.

"Fighting the Slick." *The Economist,* 9 January 1988, 25–26.

Frith, Clifford, and Dawn Frith. *Australian Tropical Reef Life.* Queensland (Australia): Frith and Frith Books, 1992.

Fujita, Rodney, et al. *A Guide to Protecting Coral Reefs.* New York: Environmental Defense Fund, 1992.

Gardiner, J. Stanley. *Coral Reefs and Atolls.* London: Macmillan and Company, 1931.

Garraty, John, and Peter Gay, eds. *The Columbia History of the World.* New York: Harper & Row, 1972.

Gato, Jeannette, ed. *The Monroe County Environmental Story.* Big Pine Key, Florida: The Monroe County Environmental Education Task Force, 1991.

Ghiselin, Michael. Foreword to Charles Darwin, *The Structure and Distribution of Coral Reefs.* Tucson: University of Arizona Press, 1984.

Ginsburg, Robert, compiler. *Proceedings of the Colloquium on Global Aspects of Coral Reefs: Health, Hazards and History.* Miami: Rosenstiel School of Marine and Atmospheric Science, University of Miami, 1994.

Gittings, Stephen. "Coral Reef Destruction at the *M/V Alec Owen Maitland* Grounding Site, Key Largo National Marine Sanctuary," Texas A&M Research Foundation, Project Number 6795, Technical Report 91-071, Texas A&M University, College Station, Texas, May 1991.

Gittings, Stephen, and Thomas Bright. "The M/V *Wellwood* Grounding: A Sanctuary Case Study." *Oceanus,* Spring 1988, 35–41.

Gleason, Bill. "The Future of Our Oceans." *Skin Diver,* August 1996, 3.

Glynn, Peter. "Coral Reef Bleaching in the 1980s and Possible Connections with Global Warming." *Trends in Ecology and Evolution* 6 (June 1991), 175–79.

———. "Extensive 'Bleaching' and Death of Reef Corals on the Pacific Coast of Panama." *Environmental Conservation* 10 (Summer 1983), 149–54.

———. "Widespread Coral Mortality and the 1982–83 El Niño Warming Event." *Environmental Conservation* 11 (Summer 1984), 133–46.

Gomez, Edgardo, et al. "Status of Philippine Coral Reefs—1981." In *Proceedings of the Fourth International Coral Reef Symposium*, vol. 1. Quezon City, Philippines: Marine Science Center, University of the Philippines, 1981, 275–82.

Goreau, Thomas. "Community Based Whole Watershed and Coastal Zone Management in Jamaica." In *Proceedings of the Eighth International Coral Reef Symposium*. In press.

Goreau, Thomas and Raymond Hayes. "Coral Bleaching and Ocean 'Hot Spots.'" *Ambio* 23, no. 3 (1994), 176–80.

Group of Experts on the Scientific Aspects of Marine Pollution. *The State of the Marine Environment*. Nairobi, Kenya: UNEP Regional Seas Reports and Studies, no. 115. United Nations Environment Programme, 1990.

Guilcher, Andre. *Coral Reef Geomorphology*. New York: John Wiley & Sons, 1988.

Harper, Peter, and Laurie Fullerton. *Philippines Handbook*. Chico, Calif.: Moon Publications, 1993.

Harrison, Peter, et al. "Mass Spawning in Tropical Reef Corals." *Science*, 16 March 1984, 1186–88.

Hay, Mark. "Seaweeds and the Ecology and Evolution of Coral Reefs." In *Proceedings of the Eighth International Coral Reef Symposium*. In press.

Hayes, Raymond, and Alan Strong. "Coral Bleaching: Ecological and Economic Impacts." Presentation to "Second Monday Seminar Program," U.S. Global Change Research Program, 13 February 1996.

Hendrickson, Robert. *The Ocean Almanac*. New York: Doubleday, 1984.

Heywood, V. H. *Global Biodiversity Assessment*. United Nations Environmental Programme. Cambridge, England: Cambridge University Press, 1995.

Hinrichsen, Don. "Reef Revival." *The Amicus Journal*, Summer 1996, 22–25.

Hodgson, Gregor, and John Dixon. "Logging Versus Fisheries and Tourism in Palawan: An Environmental and Economic Analysis." Occasional Paper No. 7. Honolulu: East–West Environment and Policy Institute, 1988.

Hopley, David. "Fringing Reefs." *Reef Notes*, March 1989. Townsville, Australia: Great Barrier Reef Marine Park Authority.

Hughes, Terence. "Catastrophes, Phase Shifts, and Large-Scale Degradation of a Caribbean Coral Reef." *Science*, 9 September 1994, 1547–51.

Humann, Paul. *Reef Creature Identification.* Jacksonville, Florida: New World Publications, 1992.

———. *Reef Fish Identification.* Jacksonville, Florida: New World Publications, 1994.

Hungspreugs, Manuwadi. "Heavy Metals and Other Non-Oil Pollutants in Southeast Asia." *Ambio* 17, no. 3 (1988), 178–182.

Hyman, L. H. *The Invertebrates.* Vol. 1 of *Protozoa through Ctenophora.* New York: McGraw-Hill, 1940.

"Indonesia: Environment and Development." Washington D.C.: World Bank, 1994.

Jaap, Walter. *The Ecology of the South Florida Coral Reefs.* U.S. Fish and Wildlife Service, U.S. Department of the Interior. Washington, D.C.: U.S. Government Printing Office, 1984.

Jackson, Jeremy. "Reefs Since Columbus." In *Proceedings of the Eighth International Coral Reef Symposium.* In press.

———. "The role of science in coral reef conservation and management." In *The International Coral Reef Initiative: Partnership Building and Framework Development.* Report of the ICRI Workshop, Dumaguete City, the Philippines, 29 May–2 June 1995, 5–9.

Jameson, Stephen, et al. "State of the Reefs: Regional and Global Perspectives." An International Coral Reef Initiative Executive Secretariat Background Paper. National Oceanographic and Atmospheric Administration, U.S. Department of Commerce. Washington, D.C.: U.S. Government Printing Office, 1995.

Johannes, Robert. "How to Kill a Coral Reef I." *Marine Pollution Bulletin* 1, no. 12 (1970), 186–87.

Johannes, Robert. "How to Kill a Coral Reef II." *Marine Pollution Bulletin* 2, no. 1 (1971), 9–10.

Johannes, Robert, and Michael Riepen. "The Environmental, Economic, and Social Implications of the Live Reef Fish Trade in Asia and the Western Pacific." The Nature Conservancy, 1995.

Kaplan, Eugene. *The Peterson Field Guide Series: Coral Reefs.* Boston: Houghton Mifflin, 1982.

Kricher, John. *A Neotropical Companion.* Princeton, N.J.: Princeton University Press, 1989.

Laber, Jeri. "Smoldering Indonesia." *The New York Review*, 9 January 1997, 40–45.

Ladd, Harry, and Seymour Schlanger. *Drilling Operations on Eniwetok Atoll.* U.S. Geological Survey, Professional Papers, 260-Y. Washington: United States Government Printing Office, 1960.

Ladd, Harry, et al. "Drilling On Eniwetok Atoll, Marshall Islands." *Bulletin of the American Association of Petroleum Geologists*, October 1953, 2257–80.

Landini, Walter, and Lorenzo Sorbini. "Ecological and trophic relationships of Eocene Monte Bolca (Pesciara) fish fauna." *Boll. Soc. Paleont. Ital.* 3 (1996), 105–12.

Lapointe, Brian, and Mark Clark. "Interim Report 1: Ambient Water Quality Assessment in the Middle and Lower Florida Keys During Summer 1989." Marathon, Florida: Florida Keys Land and Sea Trust, 1990.

Lapointe, Brian, and J. D. O'Connell. "The effects of on-site sewage disposal systems on nutrient relations of groundwaters and near-shore surface waters of the Florida Keys." Technical report, Monroe County (Florida) Planning Department, 1988.

Larkum, A., and R. West. "Long-term changes of seagrass meadows in Botany Bay, Australia," *Aquatic Botany* 37 (1990), 55–70.

Lassuy, Dennis. "Effects of 'farming' behavior by *Eupomacentrus Lividus* and *Hemiglyphidodon Plagiometopon* on algal community structure." *Bulletin of Marine Science* 30 (1980), 304–12.

Lessios, Haris. "Mass mortality of *Diadema antillarum* in the Caribbean: what have we learned?" *Annual Review of Ecological Systems* 19 (1988), 371–93.

Lessios, Haris, et al. "Spread of *Diadema* Mass Mortality Through the Caribbean." *Science* 226 (19 October 1984), 335–37.

Live Reef Fish Information Bulletin. Newsletter of the South Pacific Commission Marine Resources, Division Information Section, March 1996.

Lurie, Alison. *The Truth About Lorin Jones.* Boston: Little Brown, 1988.

Lutz, Peter, and John Musick. *The Biology of Sea Turtles.* Boca Raton, Florida: CRC Press, 1997.

MacLeish, William. *The Gulf Stream.* Boston: Houghton Mifflin, 1989.

Maragos, J. E., et al. "Coral Reefs and Biodiversity: A Critical and Threatened Relationship." *Oceanography* 9 (1996), 83–99.

McAllister, Don, and Alejandro Ansula. *Save Our Coral Reefs.* Ottawa, Canada: Ocean Voice International, 1993.

McCosker, John. "Fright Posture of the Plesiopid Fish *Calloplesiops altivelis*: An Example of Batesian Mimicry." *Science*, 22 July 1977, 400–401.

McKibben, Bill. *The End of Nature.* New York: Random House, 1989.

McManus, John. "Coral Reefs of the ASEAN Region: Status and Management." *Ambio* 17, no. 3 (1988), 189–93.

———. "Social and Economic Aspects of Reef Fisheries and their Management." In *Reef Fisheries.* Edited by Polunin and Roberts. London: Chapman & Hall, 1996, 249–81.

———. "Tropical Marine Fisheries and the Future of Coral Reefs: A Brief Review with Emphasis on Southeast Asia." In *The Proceedings of the Eighth International Coral Reef Symposium.* In press.

Meehan, William, and Gary Ostrander. "Coral Bleaching: A Potential Biomarker of Environmental Stress." *Journal of Toxicology and Environmental Health* 50 (1997), 101–24.

Moll, Hans. "Zonation and Diversity of Scleractinia on Reefs off S.W. Sulawesi, Indonesia," Ph.D dissertation, University of Leiden, the Netherlands, 1983.

Morreale, Stephen, et al. "Migration corridor for sea turtles." *Nature* 384 (28 November 1996), 319–20.

National Research Council. *Decline of the Sea Turtles.* Washington, D.C.: National Academy Press, 1990.

Odum, Eugene. *Ecology.* 1963. Reprint, New York: Holt, Rinehart and Winston, 1975.

Odum, Howard, and Eugene Odum. "Trophic structure and productivity of a windward coral reef community on Eniwetok Atoll." *Ecological Monographs* 25 (July 1955), 291–320.

Ogden, John, and Elizabeth Gladfelter, eds. "Coral reefs, seagrass beds and mangroves: Their interaction in the coastal zones of the Caribbean." *UNESCO Reports in Marine Science* 23 (1983).

Ongkosongo, Otto. "Some Harmful Stresses to the Seribu Coral Reefs, Indonesia." In *Proceedings of the MAB-COMAR Regional Workshop on Coral Reef Ecosystems.* Jakarta: UNESCO, 1986, 133–42.

Phillips, R. C., and C. P. McRoy. *Handbook of Seagrass Biology.* New York: Garland STPM Press, 1980.

Phillips, Ronald, and Ernani Meñez. "Seagrasses." *Smithsonian Contributions to the Marine Sciences* 34, 1988.

Porter, James, and Ouida Meier. "Quantification of Loss and Change in Floridian Reef Coral Populations." *American Zoologist* 32 (1992), 625–40.

Putman, R. J., and S. D. Wratten. *Principles of Ecology*. Berkeley and Los Angeles: University of California Press, 1984.

Quammen, David. *The Song of the Dodo*. New York: Scribners, 1996.

Reaka-Kudla, Marjorie, et al. *Biodiversity II*. Washington, D.C.: Joseph Henry Press, 1997.

Reef Line. (Newsletter of Reef Relief, Key West, Florida), Spring 1996.

ReefBase: A Global Database on Coral Reefs and Their Resources. CD-ROM, version 1.0. Manila, Philippines: ICLARM. June 1996.

Rhodes, Richard. *Dark Sun*. New York: Simon and Schuster, 1995.

Roberts, Callum, and Nicholas Polunin. "Marine Reserves: Simple Solutions to Managing Complex Fisheries." *Ambio* 22, no. 6 (1993), 363–68.

Roberts, Callum. "'Desconstructing' Coral Reefs." *Marine Pollution Bulletin* 28 (1994), 266.

Roessler, Carl. *The Underwater Wilderness: Life Around the Great Reefs*. New York: Chanticleer Press, 1977.

Rudloe, Jack. *Time of the Turtle*. New York: Alfred A. Knopf, 1979.

Russ, Gary. "Fisheries Management: What Chance on Coral Reefs?" *Naga*, July 1996, 5–9.

Russ, Gary, and Angel Alcala. "Sumilon Island Reserve: 20 Years of Hopes and Frustrations." *Naga*, July 1994, 8–12.

———. "Do Marine Reserves Export Adult Fish Biomass? Evidence from Apo Island, Central Philippines." *Marine Ecology Progress Series* 132, 29 February 1996, 1–9.

Sale, Peter, ed. *The Ecology of Fishes on Coral Reefs*. San Diego: Academic Press, 1991.

Salvat, Bernard. "Human Societies and Reefs, Why the Situation? Why the Challenge?" In *The International Coral Reef Initiative: Partnership Building and Framework Development*. Report of the ICRI Workshop, Dumaguete City, the Philippines, 29 May–2 June 1995, 9–13.

Samarrai, Fariss. "Helping Urchins May Benefit Corals." *Sea Frontier*, Winter 1995, 16–17.

Sammarco, Paul. "Comments on Coral Reef Regeneration, Bioerosion, Biogeography, and Chemical Ecology: Future Directions." *Journal of Experimental Marine Biology*. In press.

Sammarco, Peter, et al. "Competitive Strategies of Soft Corals (Coelenterata: Octocorallia): Allelopathic Effects on Selected Scleractinian Corals." *Coral Reefs* 1 (1983), 173–78.

Savina, Gail, and Alan White. "A Tale of Two Islands: Some Lessons for Marine Resource Management." *Environmental Conservation* 13, no. 2 (Summer 1996), 107–13.

Shepley, James, and Clay Blair Jr. *The Hydrogen Bomb*. New York: David McKay, 1954.

Singh, Someshwar. "Destructive fishing practices: Asia's growing curse." Gland, Switzerland: The World Wide Fund For Nature, 1996.

Soekarno. *Coral Reef and Associate Habitats in the Vicinity of Bunaken Togian Islands and Takabone Rate Atoll: Their Conservation Value and Needs*. Jakarta: Center for Oceanological Research and Development, 15 December 1989.

Sorokin, Yuri. *Coral Reef Ecology*. Berlin: Springer, 1993.

Steinbeck, John. *The Log from the* Sea of Cortez. 1941. Reprint, London & New York: Penguin Books, 1977.

Stimson, J. S. "Mode and Timing of Reproduction in Some Common Hermatypic corals of Hawaii and Enewetak." *Marine Biology* 48 (1978), 173–84.

Szmant, Alina, and A. Forrester. "Water column and sediment nitrogen and phosphorus distribution patterns in the Florida Keys, U.S.A." *Coral Reefs* 15 (1996), 21–41.

Szmant-Froelich, Alina, et al. "Gametogenesis and Early Development of the Temperate Coral *Astrangia Danae*." *Biology Bulletin* 158 (April 1980), 257–69.

Terres, John. *The Audubon Society Encyclopedia of North American Birds*. New York: Wings Books, 1991.

Thresher, R. E. *Reproduction in Reef Fishes*. Neptune City, N.J.: TFH Publishing, 1984.

Thurman, Harold. *Essentials of Oceanography*. Columbus, Ohio: Merrill Publishing Co., 1987.

Tomlinson, P. *The Botany of Mangroves*. Cambridge, England: Cambridge University Press, 1986.

Umbgrove, Johannes. *Madreporaria from the Bay of Batavia.* Leiden, the Netherlands: E. J. Brill, 1939.

———. "Coral Reefs of the East Indies." *Bulletin of the Geographic Society of America* 58 (August 1947), 729–77.

UNESCO. "Human Induced Damage to Coral Reefs." Reports in Marine Science, no. 40. Paris: 1986.

———. "Report on the Coral Reef Management Workshop for Pulau Seribu," Study No. 12. In *Contending with Global Change.* Jakarta: August 1996.

U.S. Congress. House of Representatives. Committee on Merchant Marine and Fisheries, Subcommittee on Oceanography and the Great Lakes and the Subcommittee on Fisheries and Wildlife Conservation and the Environment. *Hearings,* 101st Congress, 2nd session, 10 May. Washington, D.C.: U.S. Government Printing Office, 1990.

U.S. Congress. Senate. Committee on Appropriations, Subcommittee on Commerce, Justice, State, the Judiciary, and Related Agencies. "Bleaching of coral reefs in the Caribbean." *Hearings.* 100th Congress, 1st session. Washington, D.C.: U.S. Government Printing Office, 1988.

U.S. Congress. Senate. Committee on Commerce, Science, and Transportation. "Coral Bleaching." *Hearings.* 101st Congress, 2nd session. Washington, D.C.: U.S. Government Printing Office, 1991.

———. *Statutes at Large,* vol. 104, part 4. Washington, D.C.: U.S. Government Printing Office, 1991.

U.S. Department of Commerce. National Oceanic and Atmospheric Administration. "El Niño and Climate Prediction." *Reports to the Nation on Our Changing Planet.* Washington, D.C.: U.S. Government Printing Office, 1994.

———. *Draft Management Plan/Environmental Impact Statement, Vol. 1: The Management Plan.* Washington, D.C.: U.S. Government Printing Office, 1995.

———. "Diagnostic Advisory, 97/9: El Niño/Southern Oscillation (ENSO)." Climate Prediction Center, National Centers for Environmental Prediction. Washington, D.C.: U.S. Government Printing Office, 1997.

Vaccari, Ezio, and Patrick N. Wyse Jackson. "The Fossil Fishes of Bolca and the Travels in Italy of the Irish Cleric George Graydon in 1791." *Museol. Sci.* 12 (1995), 57–81.

Vannucci, Marta. "The UNDP/UNESCO Mangrove Programme in Asia and the Pacific." *Ambio* 17, no. 3 (1988), 214–17.

Vaughan, Thomas, and John Wells. "Revision of the Suborders, Families, and Genera of the Sceractinia." *Special Papers* 44, The Geological Society of America. Boulder, Colo.: 1943.

Veron, J. E. N. *Corals in Space and Time.* Ithaca, N.Y.: Cornell University Press, 1995.

———. *Corals of Australia and the Indo-Pacific.* Sydney: Angus & Robertson, 1986.

Vine, P. "Effect of Algal Grazing and Aggressive Behaviour of the Fishes *Pomacentrus lividus* and *Acanthurus sohal* on Coral-Reef Ecology." *Marine Biology* 24 (1974), 131–36.

Vogt, Helge. "The Economic Benefits of Tourism in the Marine Reserve of Apo Island, Philippines." In *The Proceedings of the Eighth International Coral Reef Symposium.* In press.

Wallace, Alfred R. *The Malay Archipelago.* London: Macmillan and Co., 1869.

Waller, Geoffrey, ed. *SeaLife: A Complete Guide to the Marine Environment.* Washington, D.C.: Smithsonian Institution Press, 1996.

Warner, Robert, et al. "Sex Change and Sexual Selection." *Science* 190 (1975), 633–38.

Warner, Robert, and Stephen Swearer. "Social Control of Sex Change in the Bluehead Wrasse, *Thalassoma bifasciatum* (Pisces: Labridae)" *The Biological Bulletin* 181 (1991), 199–204.

Wells, J. *A Survey of the Distribution of Reef Coral Genera in the Great Barrier Reef Region.* Ithaca, N.Y.: Cornell University Press, 1955.

Wells, John. "Coral growth and geochronometry." *Nature* 197 (1963), 948–50.

Wells, Susan. *Coral Reefs of the World.* Gland, Switzerland, and Cambridge, England: IUCN, and Nairobi, Kenya: UNEP, 1988.

Whitten, Anthony, et al. *The Ecology of Sulawesi.* Yogyakarta, Indonesia: Gadjah Mada University Press, 1987.

Wilkinson, Clive, ed. *Living Coastal Resources of Southeast Asia: Status and Management.* Report of the Consultative Forum, Third ASEAN-Australian Symposium. Townsville, Australia: Australian Institute of Marine Science, 1994.

Wilkinson, Clive, et al., eds. *Volume 1: Status Reviews.* Proceedings, Third ASEAN-Australian Symposium on Living Coastal Resources. Townsville, Australia: Australian Institute of Marine Science, 1994.

Willis, Bette. "Mating Systems, Hybridization and Speciation in Mass Spawning Reef Corals." In *Proceedings of the Eighth International Coral Reef Symposium.* In press.

Willoughby, N., et al. "The Effects of Human Population Pressure on Fishing Methods: From Nets to Dynamite to Cyanide." *Symposium on Environmental Aspects of Responsible Fisheries.* Seoul, South Korea, October 15–18, FAO, Asia Pacific Fisheries Commission, 1996, Section VI, 1–22.

Wilson, E. O. *The Diversity of Life.* Cambridge, Mass.: The Belknap Press of Harvard University Press, 1992.

Windhorn and Langley. *Yesterday's Key West.* Key West, Florida: Langley Press, 1973.

Wong, Marina, and Jorge Ventocilla. *A Day on Barro Colorado Island.* Panama: Smithsonian Tropical Research Station, 1995.

Yonge, C. M. *A Year on the Great Barrier Reef.* London and New York: G. P. Putnam's Sons, 1931.

Acknowledgments

First thanks must go to Alison Picard, agent extraordinaire, for keeping this project alive and finding a home for it after lo these many years. From the depths, thank you, Alison.

Many people played an important role in making this book a reality. Paul Gruchow convinced me, with encouragement and by eloquent example, that non-scientists have something to offer the field of nature writing. Connie Mutel, another non-scientist who writes about nature with passion and authority, also provided encouragement—and many helpful suggestions for the first several chapters.

Dozens of scientists were extremely helpful in providing material, making suggestions for areas to research, and walking me through the complexities of their fields. Erich Mueller, who taught Mote Marine Laboratory's summer course in coral reef ecology, is a patient and wise instructor who generously shared his thoughts via E-mail, long after the course was over. Likewise, Tony Larkum was a fount of information on many subjects and, along with his wife, Hillary, was a wonderful host in Sydney, Australia. Many thanks, too, to Steve Hendrix, at the University of Iowa, who allowed me to sit in on his ecology course several years ago, and suffered my many "questions of the day" over the past year.

Evan Edinger was not just a "source" but a friend, who made my stay in Semarang, Indonesia, productive, fascinating, and enjoyable. The same goes for Lida Pet, who set up interviews, provided a place to stay, and introduced me to the joys of *becak*s (pedicabs) and earsplitting movie theaters in Ujungpandang, Indonesia.

The following individuals each contributed to this project (though, of course, they are in no way responsible for the contents): Tim Adams, Laddie Akins, Randall Arauz, Richard Aronson, Irdez Azhar, Nenny Babo, David Bellwood, Jean-Marc

Bergevin, Maya Borel Best, Adriana Bilgray, Charles Birkeland, Rooney Biusing, Paul Blanchon, Jim Bohnsack, Charles Booth, Amanda Bourque, Massimo Boyer, Maria Majela Brenes, Gidon Bromberg, Dave Browne, Jody Bruton, Ann Budd, William Burns, Billy Causey, Herman Cesar, Milani Chaloupka, John Clark, Stephen Colwell, Sue Cook, Sergio Cotta, Marianus Dharma Datubara, Michael De Alessi, Lyndon DeVantier, Rili Djohani, Terry Done, Phillip Dustan, Ian Dutton, Mark Eakin, Nabil El-Khodari, Aaron Ellison, Jim Enright, Mark Erdmann, Daphne Fautin, Stefano Fazi, Doug Fenner, Nancy FitzSimmons, Gert Jan Gast, Jonathan Geller, Amatzia Genin, Robert Ginsburg, Edgardo Gomez, Thomas Goreau, Richard Grigg, Hector Guzman, Sudharto Hadi, Beverly Hannon, Alastair Harborne, Preston Hardison, Robin Harger, Ben Haskell, Moshira Hassan, Mark Hay, James Hendee, Selina Heppell, Gregor Hodgson, Bert Hoeksema, Anthony Hooten, Terry Hughes, Jeremy Jackson, Patrick Wyse Jackson, Stephen Jameson, Robert Johannes, Gene Kaplan, Donald Keith, Ursula Keuper-Bennett, Nancy Knowlton, Jean-Luc de Kok, Brian Lapointe, Gayatri Lilley, Don McAllister, Tim McClanahan, Kate McGill, John McManus, Alice Marlow, Marie-Trees Meereboer, Dave Meyer, Karen Miller, Willem Moka, Hans Moll, Gerardo Leyte Morales, Nyawira Muthiga, Peter Nelson, Andy Newman, Tim Norman, John Ogden, Jamie Oliver, Alfonso Aguilar-Perera, Donald Potts, Vaughan Pratt, Alfredo Quarto, DeeVon and Craig Quirolo, Pete Raines, Marjorie Reaka-Kudla, Jennifer Rendell, John Rewald, Michael Risk, Callum Roberts, Dave Robichaud, Maria Joao Rodrigues, Sandra Romano, Perran Ross, Marcy Roth, Yvonne Sadovy, Peter Sale, Rodney Salm, Paul Sammarco, Leigh Slater, Lorenzo Sorbini, Filipina Sotto, Bob Steneck, Lisman Sumardjani, Mary Swander, Alina Szmant, Frank Talbot, Dennis Taylor, Peter Thomas, Ed Tongson, Jacob van der Land, Charles Veron, Anthony Viner, Helge Peter Vogt, Sue Wells, Alan White, Ian Whittington, Simon Wilkinson, Clive Wilkinson, Nick Willoughby, Nuning Wirjoatmodjo, Gert Wöerheide, Janie Wulff.

If you're still out there, thanks to the cadre of friends with whom I first discovered the wonders of Key West more than two decades ago: Cory (King of the Conchs) McDonald, Pete Vernasco, Maryann Laraia, Peggy, Chris, Tommy, Eagle, and all the brothers and sisters of that era whose names I have forgotten, but whose spirit I'll always remember.

Thanks to my editor at Wiley, Emily Loose, who was enthusiastic about this project from start to finish.

And, finally, my greatest debt is to my family, who sacrificed so that I could follow my dream of exploring coral reefs. Thank you, once again and always.

Index

Aconcagua, 25
Acropora (acroporids)
  A. formosa, 77
  A. palmata, 44–45
  A. valenciennesi, 9
  indirect competition and, 81
  in fore-reefs, 57
  natural history, 43–44
  relocation, 79–81
  reproduction, 73, 75
  "Revolution," 44
  sea level and, 49
  skeletons, 45
  staghorn corals, 73
adaptations
  of *Acropora*, 44
  coloration, 88
  of sea grasses, 65–66
Agassiz, Alexander, 31, 34–35, 212
Age of Fishes, 39
agriculture, industrialized, 165–66
Aguadargana reef complex, 64
Air Kecil, 125
air pollution, 123
algae
  blooms and overgrowth, 70, 120, 124, 164, 182–83, 216
  control, 96–97
  coralline, 57
  green bubble, 164–65
  herbivorous fish and, 96
  red, 57
  as reef builders, 41
  rhizoselenia, 216
  zooxanthellae, 16–18, 40, 187, 188
algal ridges, 57
Amazon Cone, 161
Ameiva lizards, 169
Andes, 25–27
anemone fish, 93
angelfish, 9, 87
angiosperms, 65
anglerfish, 88–89
Animalia kingdom, 14
Anthozoa, 42
Apo Island, 153–57
archaeocyanthids, 41
Ascension Island, 108
Asia, 121, 128
Aspidontus taeniatus, 89
Astrangia danae, 76
Astrosclera, 41–42
atmosphere, 184–187, 191
atolls, 27, 29, 31–35
Atomic Energy Commission, 31
Australia, 134, 203. *See also* Great Barrier Reef
Aw, Michael, 143

Babo, Nenny, 142
back-reef, 56, 57, 60
backstepping, 49
bacteria, as reef builders, 41
Bacuit Bay, 174–75
Bahamas, 20
Bali, 53, 134, 195–96
Barang Lompo, 138
barrier reefs, 29–30
Barro Colorado, 166–69, 172
basalt, 31, 33, 34, 35

Batesian mimicry, 88
Bay of Campeche, 162
Bay of Valparaiso, 25–26
beetles, 166
Belize, 20, 192
Bell, Peter, 217
"bends," 131–32, 142
Bermuda, 19
Beston, Henry, 111–12
Bikini Atoll, 31, 32
biodiversity
  centers of, 52–54, 126–27
  of coral reefs, 5–6, 91
  decline in, 127
  of Indonesia, 126–27
  intermediate disturbances and, 178
  of ocean, 4–5
  of tropical rain forests, 166, 168–71
bioindicators, 214–15
Biology of Sea Turtles, The (Lutz and Musick), 102, 143
biomass, 170–71
bivalves, 41
black corals, 59
blast fishing, 131–36, 163
bleaching, coral, 16, 187–93
boat groundings, 207–11
Bolca fishes, 85–86
Botany Bay, 203–4
boxfish, 9
Brazil, 25, 108
brooding, 74
Brown, Barbara, 166
Browne, Janet, 26
bryozoans, 41
budding, 74
buffer zones, 70, 200–201
Bunaken, 60, 61
Bunaken Marine Park, 134
butterflies, 168–69
butterflyfish, 87–88, 95

calcification, 17–18
calcium carbonate, 17–18, 22, 43, 57
Calloplesiops altivelis, 88
camouflage, 87–89
Canada, 157
Canadian Maritimes, 4
carbon dioxide, 191
Caribbean Sea, 53, 120, 180–83, 190
Carson, Rachel, 4, 59, 68, 144, 175
catastrophes, 179, 183
catch-up reefs, 49
Causey, Billy, 218–20
centinelan extinctions, 173–74
Chelonia mydas. See green sea turtle
Chile, 25–26
Chordata, 61–62
Christmas Island, 186
Christmas tree worms, 64
chromatophores, 10
clams, giant, 28
cleaner fishes, 89
cleaning symbiosis, 89
climate, 21–22, 48
Clinton, Bill, 222
clownfish, 9
Cnidaria, 42
Cocos-Keeling Islands, 28–29
coelenterates, 42
colonies
  formation, 79
  structure, 14
color
  changes, 10, 93
  initial phase (IP), 92
  of reef fishes, 86–94
  in reefs, 58
  and sexual behavior, 89–94
  terminal phase (TP), 91–92
Columbus, Christopher, 184
comb jellies, 108
comet fish, 88
competition, 81–82, 183
Concrete Coalition, 213
cone shells, 8
Connell, Joseph, 178
continental drift, 48

Cook, James, 79–80
coralline sponges, 41–42
corallite, 14, 17
coral polyps
  anatomy, 14–15
  axial, 45
  radial, 45
coral reefs
  barrier, 29–30
  biodiversity in, 5–6, 91
  catch-up, 49
  as climatic recorders, 21–22
  construction, 17–18, 40–42, 57
  crests, 56, 57, 60
  deep-water, 150
  destruction, 11, 120, 124–28, 131–36 (*see also* human threats to reefs)
  distribution of, 18–20
  ecology of, 96–98
  effects of collapse, 127
  environmental requirements, 18–20, 189–92
  erosion, 18, 66–67
  as food source, 21
  formation theory, 23–31, 34–35
  fringing, 29–30
  future of, 221–22
  give-up, 49
  as habitats, 22
  importance of, 21–22
  interactions among, 81
  keep-up, 49
  mangroves and, 67, 70
  misconceptions about, 46, 47–48
  sea grasses and, 67, 70
  sizes of, 18
  total area of, 5, 20–21
  tropical rain forests and, 6, 172–74
  windward, 58
  worldwide status of, 127–28
  zones, 56, 57
corals. *See also* coral polyps; coral reefs; *specific kinds*
  as "animal-plant-mineral," 18
  calcification of, 17–18
  colonial, 40, 43
  diseases, 212, 216
  food requirements, 15–16
  hard, 42, 43, 58
  natural history, 8, 39–43, 45–50
  natural threats to, 45–46, 49–50, 81–83
  predators of, 95
  reef fishes, 95
  soft, 43, 58, 59
  taxonomy of, 14
  water temperature and, 189–92
  zooxanthellae and, 16–18, 40, 187, 188
coral trout, 138
Coralville Dam, 38–39
Cotta, Sergio, 51, 57–58, 60
creationism, 27
Cretaceous crash, 105
Cretaceous period, 65
crinoids, 59
Crossland, Chris, 152
crown-of-thorns sea stars, 188–89
crustaceans, "guardian," 189
currents, 79–81, 161–62, 185
cushion stars, 8
cyanide fishing, 140–43, 149
cyanobacteria, 41

damselfish, 97
Dana, James, 18–19
Darwin, Charles
  circumnavigation of globe by, 23–29
  on Cocos-Keeling Islands, 28–29
  coral reef descriptions by, 7, 18, 27
  illness of, 26
  and mangroves, 67–68
  in rain forests, 25
  subsidence theory, 30, 34, 48
  theory of evolution, 54
  theory of reef formation, 23–31, 34–35

decompression sickness, 131–32, 142
deforestation, 172
developers, 213
Devonian crash, 40
Devonian period, 39
Diadema
  D. antillarum, 180–83
  die-off, 180–83, 188, 215
  in Jakarta Bay, 120
Dictyosphaeria cavernosa, 164–65
dinosaurs, 105, 179
disasters, 178–79
Dixon, John, 173
dredging, 126
drilling, 31–35
dugongs, 201–2

Earle, Sylvia, 144
Echinodermata (echinoderms), 59
ecological preservation zones, 220–21
ecology, 55, 96–98
Ecology of Fishes on Coral Reefs, The (Sale), 47
economics. *See* socioeconomics of coral reefs
ecosystems
  closed, 94
  coral reefs, 55–56, 63, 67
  heterotrophic, 63
  linked, 67, 70
  mass die-offs in, 181–83
  open, 63–64, 81, 95
  redundant species in, 184
  species relationships in, 181–83
Edinger, Evan, 135, 163, 198, 199
eels, 8, 88
elkhorn corals, 44–45
El Niño. *See* ENSO events
Elugelab, 32–34, 35, 36
End of Nature, The (McKibben), 193
Enewetak Atoll, 31–32, 34–36, 55
ENSO (El Niño Southern Oscillation) events, 185–91
Eocene Epoch, 44, 86
equator, 52
Erdmann, Mark, 133, 135
erosion, 18, 66–67, 172, 174
eutrophication, 120, 164, 217
evolution by natural selection, theory of, 6, 30, 53–54
extinction
  centinelan, 173–75
  "debt," 82–83
  of dinosaurs, 105, 179
  events, 40, 127, 179
  of islands, 126
  of passenger pigeons, 179
  periodic mass, 45–46, 49
  of reef fishes, 142
Exxon Valdez, 210

false eye-spots, 87–88, 169
Fascell, Dante, 210, 211
Fazi, Stefano, 115–18, 128
Fiji, 34–35
filter feeders, 61
fisheries management, 151, 157
fishes, reef. *See also specific kinds*
  adaptations, 88
  algae consumption by, 96
  body types, 86–87
  camouflage, 87–89
  coloration, 86–94
  corals and, 95
  extinction of, 142
  in Great Barrier Reef, 8–11
  herbivorous, 96
  as "immune system," 96
  larval stage, 94
  live catches, 138–43, 149
  numbers of, 5, 86
  in open water, 94
  sea-grass beds and, 202
  sex changes in, 91–94
  sexual behavior, 89–94
fishing practices, destructive

blast fishing, 131–36, 163
cyanide fishing, 140–43, 149
live-reef-fish trade, 138–43, 149
muro-ami, 149, 153
overfishing, 120, 126, 150–51, 175
trawling, 149–50
FitzRoy, Captain, 25
"flood of the century" (1993), 37–39
Florida Bay, 215–16, 223
Florida Current, 162
Florida Keys
economic benefits of reefs to, 21
mangrove forests surrounding, 68
water quality, 20
Florida Keys National Marine Sanctuary, 209–11, 218–21
Florida reef tract
boat groundings on, 208–11
decline of, 211–15, 219
Hawk Channel, 162, 213
management plan, 219–221
residents' attitudes, 218–19
foraminiferans, 41
fore-reef, 56, 57
Fortes, Miguel, 201, 204
fossil fuels, 191–93
fossils, 26, 39–40, 85–86
French Frigate Shoals, 79, 80
fringing reefs, 29–30
Funafuti, 31
Fungia corals, 9

Galápagos Islands, 19, 187, 188
Gardiner, J. Stanley, 6, 7
give-up reefs, 49
glaciation, 48
glaciers
"glacial mode," 48
melting events, 48, 49
Wisconsin, 49
Global Coral Reef Alliance, 192
global warming, 191–93
Glynn, Peter, 187–90
"Golden Triangle" of Jakarta, 122
Gondwana, 67
Goniastrea aspera, 77
Goniastrea favulus, 77
gonorchism, 90
Goreau, Thomas, 175, 192
gorgonian corals, 43
gorgonin, 43
Graham, Bob, 210–11
Grand Banks, 157
Grant, Robert, 24
Great Barrier Reef
coral bleaching on, 190
Heron Island, 7–11, 60, 99–102, 110, 111
lagoon, 217
marine life on, 8–11
spawning events on, 78
green bubble algae, 164–65
greenhouse effect, 191–93
green sea turtles
in Caribbean, 183–84
diet, 108–9
hatchlings, 107–8
at Heron Island, 8
human threats to, 112
importance to reefs, 102, 109
juveniles, 108–9
as keystone species, 102
life span, 102
light and, 103, 107
"lost decade," 102, 108–9
mass slaughter of, 183–84
migration of, 108, 109–10
natural history, 105
nesting process, 100–1, 103–8, 110–11
predators of, 107
sea grasses and, 109, 202
sexual behavior, 109–10
Grenard, Steve, 102
Grigg, Richard, 79–81
Guiana Current, 161

Gulf of Mexico, 162
Gulf Stream, 4, 19, 20, 162
gyre, 162

hamlet fish, 93
hard corals, 42, 43, 58
hatchling frenzy, 107–8
Hawaiian Islands, 79–81, 163–66
Hawk Channel, 162, 213
heavy metals, 123–24
Hemingway, Ernest, 160
herbivory, 96
hermaphrodites, 91, 93
Heron Island, 7–11, 60, 99–102, 110, 111
Hertel, Dennis, 210
Hexacorallia (hexacorals), 43
Hickson, Sidney, 54
HMS *Beagle*, 23–29
Hodgson, Gregor, 173–75, 222
horn corals, 39–40, 46
horn protein, 59
Hughes, Terence, 183
human threats to reefs
  blast fishing, 131–36, 163
  cyanide fishing, 140–43, 149
  development, 213
  dredging, 126
  global warming, 191–93
  industrialized agriculture, 165–66
  live-reef-fish trade, 138–43, 149
  mining, 125
  overfishing, 120, 126, 150–51, 175
  pesticides, 125
  pollution, 123–24, 163–64, 212–13
  shrimp farming, 197–99
  trawling, 149–50
humphead wrasse, 142
hybridization, 78–79
hybrid zones, 78
hydrogen bomb tests, 35–36

Indonesia
  archipelago, 6–7
  biodiversity of, 126–27
  blast fishing in, 131–36
  corruption in, 137
  live-reef fish trade, 141
  reef destruction in, 120, 124–28, 131–36
Indonesian Low, 185
initial phase (IP) coloration, 92
interdisciplinarianism, 152
intermediate disturbance hypothesis, 178
International Center for Living Aquatic Resources Management (ICLARM), 147
International Coral Reef Symposiums, 49, 152, 221
International Year of the Reef, 222
Iowa River, 38–39
island extinctions, 126

Jaap, Walter, 212–13
Jackson, Jeremy, 102, 183
Jakarta, 116, 121–25, 128
Jakarta Bay, 115, 120–21, 123–27
Jamaica, 182–83
James Cook University, 73, 75–78
Java, 115, 121
Java Sea, 129
jellyfish, 42
jet stream, 186
Johannes, Robert, 164, 165
Johnston Island, 80
Jurassic period, 41

Kaneohe Bay, 163–166, 175, 214
Karimunjawa, 125
keep-up reefs, 49
Kemp's ridley sea turtle, 112
Kepulauan Sankarang, 132–33
Kepulauan Seribu, 116
Key Largo National Marine Sanctuary, 209
keystone species, 102
Key West

author in, 159–61
boat groundings in, 207–11
Kita Daito Jima, 31
Krakatau, 177–79
Kurnell National Park, 203

Ladd, Harry, 32–35
Lake Gatún, 167
Lapointe, Brian, 211–18
Larkum, Tony, 64–67, 202, 217–18
larvae, coral
recruitment, 80–81, 82
zooxanthellae and, 17
larvae, fish, 94
Laurasia, 67
leafcutter ants, 168
leatherback turtles, 108
Leiden, 119. *See also* Nyamuk Besar
lek, 91–94
lenticels, 70
Likuan, 61–62
limestone, 17–18, 33–34
Linnaeus, 13–14
live-reef-fish trade, 138–43, 149
logging, 173–75
Looe Key, 212, 218
Loop Current, 162
Lurie, Alison, 68
Luzon, 52
Lyell, Charles, 26–27, 29, 30–31

Makassar Straits, 133
Makati City, 148
Malthusian overfishing, 150
Manado, 51–53, 134
mangroves, 67–70, 197–99
"map sense," 108, 109-10
Mariana Trench, 4
marine reserves, 154–57, 174–75
Márquez, Gabriel Garciá, 193
Marshall Islands, 31
Mauritius, 30
McKibben, Bill, 193
McManus, John, 147–53, 200
medical uses of corals, 21
megacities, 121–23, 128
Mentawai islands, 134
metapopulations, 95
microatoll, 61
microclimate, 171
migration of sea turtles, 108, 109–10
mining, 125
Miocene Epoch, 48
Molasses Reef, 208
monkeys, 167–68, 169
Moorea, 27
moray eels, 88
morphotypes, 44, 86–87
Mount Pinatubo, 192
Mueller, Erich, 44, 178, 212
Munro, Mary, 207
muro-ami, 149, 153
mutualistic symbiosis (mutualism), 17

Nalunega Island, 64–65
Narragansett Bay, 74
natural disasters, 177–79
natural selection, 6, 30, 53–54, 88
nematocysts, 82
New Caledonian sea star, 8
nominal species, 44
North Pacific Equatorial Current (NPEC), 80–81, 82
notochords, 61–62
nuclear weapons testing program, 31, 35–36
nutrient cycling, 63–64, 168, 171–72
nutrient intake, 15–16
nutrient loading, 120, 124, 163–64, 212–17
Nyamuk Besar, 115–20, 126

Oahu, 164, 165
Oakley, Steve, 143
ocean
atmosphere and, 184–87
biodiversity in, 4–5
calcium sedimentation in, 17

ocean *(continued)*
  currents, 79–81, 185
  importance of reefs to, 22
  level changes in, 48
  natural history, 4–5
  transgression of, 49
  vastness, 4
Octocorallia (octocorals), 43
octopuses, 10
Odum, Eugene, 55, 63
Odum, Howard, 55
Ogden, John, 211
Oliver, Jamie, 73–75
Onychorphora, 4
oocytes, 75
overfishing, 120, 126, 150–51, 175

Palawan, 173–75
Panama Canal, 46, 167, 172
Pangaea, 39
Panthalassa, 39
Papua New Guinea, 52
Paraponera clavata, 168
parasitic symbiosis, 17
parrotfishes
  algae control by, 18, 95, 97
  in Great Barrier Reef, 9
passenger pigeons, 179
Permian crash, 40, 179
pesticides, 125
Pet, Lida, 129–33, 135–38, 140–41
Peyssonnel, 14
Phanerozoic eon, 41
pharyngeal mill, 9, 97
Philippines
  Apo Island, 153–57
  archipelago, 153
  blast fishing in, 134, 136
  cyanide fishing in, 149
  reef destruction in, 128, 134
  and triangle of biodiversity, 52
  tropical rain forests in, 173–75
photic zone, 5
phytoplankton, 20
piscivores, 89
planulae, 17
planulation, 74
Plectropomus laevis, 138
pneumatophores, 70
Pocillopora, 189
poison-arrow frog, 169–70
polar ice caps, 48, 49
pollution
  air, 123
  in shrimp ponds, 199–202
  water, 123–24, 163–64, 199–202, 212–13
polyps. *See* coral polyps
pomacentrids, 97
ponds, shrimp, 198–202, 205–6
population growth, human, 121–23, 127–28, 143, 150, 183
Porifera, 24. *See also* sponges
Porites coral, 61
pororoca, 161
Port Botany, 203
Porter, James, 212
Posidonia australis, 203–4
poverty, 143
preservation zones, 220–21
primary production, 14, 63
Principles of Geology (Lyell), 26
protogynous fishes, 93
protozoans, 59
Pseudochromis paccagnellae, 58
Punta Galeta, 181

queen angelfish, 87
queen triggerfish, 183
Quirolo, Craig, 211
Quirolo, DeeVon, 220

rafting, 81
Raine Island, 110
rain forests. *See* tropical rain forests
Rakata, 177–78

recruitment, coral, 80–81, 82
redundant species, 184
reef builders, 40–42
Reef Check, 222
Reef Creature Identification (Humann), 180
reef crest, 56, 57, 60
reef flats, 60–61
Reef Relief, 211, 220
reefs. *See* coral reefs
reproduction
  coloration and, 89–94
  corals, 74–79, 81
  lek, 91–94
  reef fishes, 89–94
  sea grasses, 66
rhizoselenia, 216
Risk, Mike, 127
Ross, Sir John, 5
Royal Society (of London), 31, 35
Rugosa corals, 40
Russ, Gary, 157

saddleback coral grouper, 138
Sale, Peter, 46–47, 50
Salm, Rodney, 134
Salvat, Bernard, 152, 221
San Blas Islands, 64, 172, 181
sand production, 97–98
scientific diversity, 151–52
Scleractinia (scleractinian corals), 42, 43, 78
sea anemones, 9, 42, 43
sea cucumbers, 8
sea fans, 59
sea grasses
  adaptations, 65–66
  coral reefs and, 67, 70
  destruction of, 202–204
  die-offs of, 215, 216
  green sea turtles and, 109
  importance of, 66–67, 201–2
  mangroves and, 70
  meadows, 64, 65–67
  natural history, 65
  reproduction, 66
sea horses, 90–91, 202
sea level, changes in, 48
sea lilies, 59
sea squirts, 61
sea stars, 8, 188–89
sea turtles. *See also* green sea turtles
  decline of, 143
  Kemp's ridley, 112
sea urchins, 120. *See also Diadema*
sea whips, 88
sedimentation, 124–25, 156, 163, 172–74
sewage, 124, 163–64, 213–15, 217
sexual plasticity, 91–94
Shark Bay, 100, 103
sharks, reef, 9
shrimp farming, 197–99
Silliman University, 154
skeletons, coral, 14, 17, 30, 40, 43, 45
Slater, Leigh, 101–7, 110–12
sneaking, 92
social facilitation hypothesis, 110
socioeconomics of coral reefs, 143–45, 150, 155, 174–75, 198
soft corals, 43, 58, 59
Sotto, Filipina, 156
Southeast Asia, 20, 127–28, 197, 200
South Equatorial Current, 161
Southern Oscillation, 186. *See also* ENSO events
spawning, 75–79, 81, 90–92
Spermonde archipelago, 132–33, 135, 142
spicules, 41
spillover effect, 155–56
spiral wire corals, 58–59
sponges, 18, 24, 41–42, 59–60, 216
spur-and-groove formations, 58
staghorn corals, 73
stone crabs, 159–60

stony corals. *See* hard corals
streaking, 92
stromatolites, 41
stromatoporoids, 41–42
Structure and Distribution of Coral Reefs, The (Darwin), 23–24
subsidence theory, 30, 34, 48
Subtropical Countercurrent (SCC), 80–81
Suharto, 122, 123, 137
Sulawesi Island, 51–53, 129–30
Sulawesi Natural Resources Conservation Information Center, 142
Sumatra, 52
sunlight
  and reef growth, 19
  and zonation, 56
superponds, 198–202, 206
sustainable use, 154–55
Swearer, Stephen, 93
symbiosis, 16–18, 40, 89, 189
Szmant, Alina, 75–76, 78, 182, 214–15

Tabulata corals, 40
Tahiti, 27
*tambak*s, 198–201, 205–6
Tanjung Pisok, 53, 54, 56–58, 61
Tanzania, 134
taxonomy, 13–14, 42–44, 69
Teller, Edward, 36
temperature, water, 18–19, 189–92
tentacles, 14, 40, 43, 82
terminal phase (TP) coloration, 91–92
terns, noody, 99
Testudines, 109
Tethys Sea, 67, 86
Thailand, 128
Thalassia, 65–67. *See also* sea grasses
  testudinum, 65, 109
theory of evolution by natural selection, 6, 30, 53–54
theory of reef formation, 23–31, 34–35
Thousand Islands, 116
tidal bore, 161
tidal waves, 200–1
trade winds, 185
tragedies, 179–80, 83
transgression, 49
trawling, 149–50
trilobites, 41
tropical rain forests
  biodiversity of, 166, 168–71
  biomass in, 170–72
  coral reefs and, 172–74
  Darwin in, 25
  destruction of, 166, 172–74
  nutrient cycling in, 168, 171–72
  similarities to coral reefs, 6
  soil erosion in, 172
  zonation in, 171
tropics, 51–52
trumpetfish, 88
tunicates, 61, 133
turtle grass, 109
turtles. *See* green sea turtles; leatherback turtles

Ujungpandang, 130, 135–37, 141
Umbgrove, Johannes, 117, 119, 120, 125, 134
U.S. Senate, 190, 192
Uva Island, 187–89

Vader Smit reef, 125
Vine, P., 96
Viner, Anthony, 199
viviparity, 74
volcanos, 27, 30, 31, 35, 177–79

Wallace, Alfred Russel, 6–7, 53–54
Warner, Robert, 93
water
  eutrophic, 20, 164, 217
  nutrients in, 20
  oligotrophic, 20
  pollution, 123–24, 163–64, 212–13

quality, 19–20
temperature, 18–19, 189–92
Wicklund, Robert, 190–91
Williams Jr., Ernest, 192
Williams, Lucy Bunkley-, 191
Willis, Bette, 78
Wilson, E. O., 166, 173, 178
Wisconsin glacier, 49
Wörheide, Gert, 41–42
worms, 5, 18, 64
wrasses
bluehead, 91–94
in Great Barrier Reef, 9
humphead, 142
reproduction, 91–94

Yucatán Channel, 162

zonation, 56, 57, 70, 171
zoning in marine parks, 220–21
zoophytes, 14, 18, 24
zooplankton, 15
zooxanthellae, 16–18, 40, 187, 188